I0789938

2011 International Conference on Semiconductor Technology for Ultra Large Scale Integrated Circuits and Thin Film Transistors (ULSIC vs. TFT)

Editors:

Y. Kuo
Texas A&M University
College Station, Texas, USA

G. Bersuker
Sematech
Austin, Texas, USA

Sponsoring Division:

 Electronics and Photonics

Published by
The Electrochemical Society
65 South Main Street, Building D
Pennington, NJ 08534-2839, USA
tel 609 737 1902
fax 609 737 2743
www.electrochem.org

ecstransactions ™

Vol. 37, No. 1

Copyright 2011 by The Electrochemical Society.
All rights reserved.

This book has been registered with Copyright Clearance Center.
For further information, please contact the Copyright Clearance Center,
Salem, Massachusetts.

Published by:

The Electrochemical Society
65 South Main Street
Pennington, New Jersey 08534-2839, USA

Telephone 609.737.1902
Fax 609.737.2743
e-mail: ecs@electrochem.org
Web: www.electrochem.org

ISSN 1938-6737 (online)
ISSN 1938-5862 (print)
ISSN 2151-2051 (cd-rom)

ISBN 978-1-56677-899-2 (CD-ROM)
ISBN 978-1-60768-251-6 (PDF)
ISBN 978-1-60768-252-3 (Softcover)

Preface

This issue of ECS *Transactions* includes 25 papers presented at the Third International Conference on Semiconductor Technology for Ultra Large Integrated Circuits and Thin Film Transistors (ULSIC vs. TFT III), held at the Hyatt Regency Hong Kong, Sha Tin, in Hong Kong, China, from June 27 to July 1, 2011. This symposium was sponsored by the Engineering Conferences International.

This conference provided a forum on the latest developments in the two largest semiconductor technologies, ULSIC and TFT. Global experts from universities and industry exchanged experience, knowledge, and visions through technical presentations and discussions during presentations, brainstorming sections, and afternoon and evening breaks. Future scientific and technology challenges were forecasted and debated. Graduate students and junior researchers from various countries also presented papers and were engaged in discussions; participants were from Belgium, China, France, Italy, Hong Kong, Japan, Korea, the Netherlands, Taiwan, UK, and the USA.

Also included at the conference were three plenary speeches: they were dedicated to 3D IC challenges by R. Jammy of Sematech; another was on semiconductor technology evolution by H.-K. Kang of Samsung; and lastly was discussed silicide nanowires by L. J. Chen of National Tsing Hua University. In total, 34 papers were presented in the following sessions:

- Plenary
- Nano Devices
- IC Devices & Materials
- Oxide TFTs
- Memories and Interfaces
- Si-based TFTs
- IC beyond Si
- TFT Devices and Processes
- Packages and New Device Principles
- TFT New Applications
- Posters

The conference also included three panel sessions on challenges in 1) extremely large and small scalings, 2) materials and fabrications, and 3) applications.

In order to present subjects in a coherent manner, papers in this issue of *ECS Transactions* have been arranged into six chapters. All manuscripts are published as originally received, without alteration of technical contents, except minor formatting corrections.

iii

The success of this conference is contributed by the following people and organizations:

- Plenary, keynote, and invited speakers for presentations and panel discussions
- Session chairs for conducting the meeting
- Authors and presenters for their participation
- Scientific advisory committee members for their contributions in planning and coordinating the program
- Semiconductor Energy Laboratory, Electrochemical Society Electronics and Photonics Division, and Sematech for financial support
- Dr. N. Li for valuable suggestions and advice as the conference ECI liaison
- Ms. Sylvia Kwok for arrangement of the student accommodation
- Dr. Anthony H. W. Choi for coordinating with IEEE ED/SSC Hong Kong
- ECI staff for effective management of the conference
- J. Lewis of the Electrochemical Society for his professional work in publishing this issue of ECS *Transactions*

Yue Kuo,
Texas A&M University

Gennadi Bersuker,
Sematech

Acknowledgements

For the generous sponsorship:

Semiconductor Energy Laboratory

The Electrochemical Society Electronics and Photonics Division

Sematech

For valuable technical sponsorship:

The Electrochemical Society Electronics and Photonics Division

Japan Society of Applied Physics

Korean Physical Society Semiconductor Division

IEEE ED/SSC Hong Kong

Sematech

imec

Conference Co-chairs
Yue Kuo, *Texas A&M University*
Gennadi Bersuker, *Sematech*

Local Committee
F. Yan, *Hong Kong Polytechnic University*

Scientific Advisory Committee

T. Asano, *Kyushu University*
O. Bonnaud, *University of Rennes I*
C. Claeys, *imec*
S. Fonash, *Pennsylvania State University*
M. Hatano, *Tokyo Institute of Technology*
J. Jang, *Kyung Hee University*
W. I. Milne, *Cambridge University*
A. Nathan, *University College London*
R. Street, *PARC*
S. Wagner, *Princeton University*

D. N. Buckley, *University of Limerick*
G. Fortunato, *CNR-IMM*
R. Ishihara, *Delft University of Technology*
T. P. Ma, *Yale University*
J. Murota, *Tohoku University*
M. Shur, *RPI*
S. Uchikoga, *Toshiba*
C.P. Wong, *Chinese University of Hong Kong & Georgia Institute of Technology*
F. Yan, *Hong Kong Polytechnic University*

Session Chairs

T. Asano, *Kyushu University*
G. Bersuker, *Sematech*
O. Bonnaud, *University of Rennes I*
S. Fonash, *Pennsylvania State University*
G. Fortunato, *CNR-IMM*
R. Ishihara, *Delft University of Technology*
J. Jang, *Kyung Hee University*
Y. Kuo, *Texas A&M University*
H. S. Kwok, *The Hong Kong University of Science & Technology*
L. Larcher, *Università di Modena e Reggio*

T. P. Ma, *Yale University*
C. Matty, *imec*
W. Milne, *Cambridge University*
J. Murota, *Tohoku University*
Y. Saito, *Toshiba*
S. Sato, *AIST*
A. Shluger, *University College London*
S. Srivanasan, *Applied Materials*
C. P. Wong, *Chinese University of Hong Kong & Georgia Institute of Technology*
F. Yan, *Hong Kong Polytechnic University*

ECS Transactions, Volume 37, Issue 1
2011 International Conference on Semiconductor Technology for Ultra Large Scale Integrated
Circuits and Thin Film Transistors (ULSIC vs. TFT)

Table of Contents

Preface   *iii*

Chapter 1
Challenges in Si- and Ge-based TFT Technology

(Invited) Short Channel Effects and Drain Field Relief Architectures in Polysilicon TFTs   3
  G. Fortunato, M. Cuscunà, L. Maiolo, L. Mariucci, M. Rapisarda, A. Pecora,
  A. Valletta, and S. D. Brotherton

(Invited) Self-Heating Issue of Poly-Si TFT on Glass Substrate   15
  T. Asano and G. Nakagawa

(Invited) Bridged-grain (BG) Polycrystalline Silicon Thin Film Transistors (TFTs)   23
  H. Kwok, M. Wong, S. Zhao, and W. Zhou

(Invited) Vertical Channel Thin Film Transistor: Improvement Approach Similar to   29
Multigate Monolithic CMOS Technology
  O. Bonnaud, P. Zhang, E. Jacques, and R. Rogel

(Invited) Similarities between µc-Si TFT with Very Thin Active Layer and FD-SOI FETs   39
  T. Mohammed-Brahim, K. Kandoussi, K. Belarbi, H. Lhermite, N. Coulon, and
  C. Simon

MILC PMOS Poly-Silicon TFT Circuits and Application in SOP   49
  P. Sun, S. Zhao, T. Ho, Z. Meng, M. Wong, and H. Kwok

Source-Gated Transistors for Versatile Large Area Electronic Circuit Design and   57
Fabrication
  R. A. Sporea, X. Guo, J. M. Shannon, and S. R. Silva

(Invited) Single-Grain Germanium TFTs   65
  R. Ishihara, T. Chen, A. Baiano, M. R. Mofrad, and C. Beenakker

vii

Chapter 2
Challenges in Oxide and Organic TFTs

(Invited) Success in Measurement the Lowest Off-state Current of Transistor in the   77
World
    Y. Sekine, K. Furutani, Y. Shionoiri, K. Kato, J. Koyama, and S. Yamazaki

Low Power 6.0-Inch Extended Graphics Array Reflective Liquid Crystal Display using   89
Indium Gallium Zinc Oxide Semiconductor with Electronic Paper Function
    M. Kaneyasu, H. Miyake, T. Nishi, Y. Hirakata, J. Koyama, R. Sato, M. Sakakura, and
    S. Yamazaki

Low Power 6.0-inch Extended Graphics Array Transmissive Liquid Crystal Display using   97
Indium Gallium Zinc Oxide Semiconductor with Variable Frame Frequency
    S. Amano, H. Miyake, T. Nishi, Y. Hirakata, J. Koyama, K. Okazaki, M. Sakakura, and
    S. Yamazaki

Comparative Analysis of Organic Thin Film Transistor Structures for Flexible E-Paper   105
and AMOLED Displays
    L. Feng, X. Xu, and X. Guo

Chapter 3
Challenges in Graphene, Nanowire, and Nanotube Devices

(Keynote) Recent Progress in Facile Preparation of Graphene   115
    Z. Lin, Y. Yao, Z. Li, K. Moon, and C. P. Wong

(Invited) Application of Graphene and Carbon Nanotubes to Transistors and   121
Interconnects
    S. Sato, K. Hayashi, K. Yagi, D. Kondo, A. Yamada, N. Harada, M. Nihei, and
    N. Yokoyama

Manipulation of Graphene Properties by Interface Engineering   133
    X. Wang, J. Xu, C. Wang, J. Du, and W. Xie

(Invited) Silicon Grow-in-Place Nanowires and Their Applications   141
    J. Wu, P. Garg, S. Pan, C. Winter, D. Scott, and S. Fonash

(Invited) In-plane Silicon Nanowires for Field Effect Transistor Application   147
    L. Yu and P. Roca i Cabarrocas

Chapter 4
Challenges in Memories

(Invited) Nonvolatile Memories for Nano and Giga Electronics   157
   Y. Kuo

(Invited) Understanding the Switching Mechanism in RRAM Devices and the Dielectric   167
Breakdown of Ultrathin High-k Gate Stacks from First Principles Calculations
   B. Magyari-Köpe, S. Park, H. Lee, and Y. Nishi

Chapter 5
Challenges in Process Engineering

(Invited) Atomically Controlled CVD Processing for Doping in Future Si-Based Devices   181
   J. Murota, M. Sakuraba, and B. Tillack

(Invited) Physical Modeling of Charge Transport and Degradation in HfO_2 Stacks for   189
Logic Device and Memory Applications
   L. Larcher, A. Padovani, L. Vandelli, and G. Bersuker

Formation and Characterization of NiYb Silicides   199
   Y. Yuan and E. Ivanov

Chapter 6
Challenges in New Devices and Applications

(Keynote) Unipolar CMOS Logic for Post-Si ULSI and TFT Technologies   207
   T. Ma and X. Sun

(Invited) Spin-based MOSFET and Its Applications   217
   Y. Saito

(Invited) Flexible Thin Film Transistor Arrays as an Enabling Platform Technology: 229
Opportunities and Challenges
   G. B. Raupp

Author Index   241

Facts about ECS

The Electrochemical Society (ECS) is an international, nonprofit, scientific, educational organization founded for the advancement of the theory and practice of electrochemistry, electrothermics, electronics, and allied subjects. The Society was founded in Philadelphia in 1902 and incorporated in 1930. There are currently over 7,000 scientists and engineers from more than 70 countries who hold individual membership; the Society is also supported by more than 100 corporations through Corporate Memberships.

The technical activities of the Society are carried on by Divisions. Sections of the Society have been organized in a number of cities and regions. Major international meetings of the Society are held in the spring and fall of each year. At these meetings, the Divisions and Groups hold general sessions and sponsor symposia on specialized subjects.

The Society has an active publications program that includes the following.

Journal of The Electrochemical Society — JES is the peer-reviewed leader in the field of electrochemical and solid-state science and technology. Articles are posted online as soon as they become available for publication. This archival journal is also available in a paper edition, published monthly following electronic publication.

Electrochemical and Solid-State Letters — ESL is the first and only rapid-publication electronic journal covering the same technical areas as JES. Articles are posted online as soon as they become available for publication. This peer-reviewed, archival journal is also available in a paper edition, published monthly following electronic publication. It is a joint publication of ECS and the IEEE Electron Devices Society.

Interface — *Interface* is ECS's quarterly news magazine. It provides a forum for the lively exchange of ideas and news among members of ECS and the international scientific community at large. Published online (with free access to all) and in paper, issues highlight special features on the state of electrochemical and solid-state science and technology. The paper edition is automatically sent to all ECS members.

Meeting Abstracts (formerly Extended Abstracts) — Abstracts of the technical papers presented at the spring and fall meetings of the Society are published on CD-ROM.

ECS Transactions — This online database provides access to full-text articles presented at ECS and ECS-sponsored meetings. Content is available through individual articles, or as collections of articles representing entire symposia.

Monograph Volumes — The Society sponsors the publication of hardbound monograph volumes, which provide authoritative accounts of specific topics in electrochemistry, solid-state science, and related disciplines.

For more information on these and other Society activities, visit the ECS website:

www.electrochem.org

CHAPTER 1

CHALLENGES IN Si- AND Ge-BASED TFT TECHNOLOGY

2

Short channel effects and drain field relief architectures in polysilicon TFTs

G. Fortunato[a], M. Cuscunà[a], L. Maiolo[a], L. Mariucci[a], M. Rapisarda[a], A. Pecora[a], A. Valletta[a] and S.D. Brotherton[b]

[a] IMM-CNR, Via del Fosso del Cavaliere, 100 - 00133 Rome, ITALY
[b] TFT Consultant, 12 Riverside, Forest Row RH18 5HB, United Kingdom

> Applications of polycrystalline silicon (polysilicon) thin film transistors (TFTs) to active matrix organic light emitting displays require further performance improvement. The biggest leverage in circuit performance can be obtained by reducing channel length from the typical current values of 3-6µm to 1µm, or less. However, short channel effects and hot-carrier induced instability in scaled down conventional self-aligned polysilicon TFTs can substantially degrade the device characteristics. To reduce these effects and allow proper operation of the circuits, drain field relief architectures have to be introduced. In this work we show that a fully self-aligned gate overlapped lightly doped drain (LDD) structure, with submicron LDD regions, can provide an excellent solution, allowing effective short channel effect control and improved electrical stability.

Introduction

Low temperature polycrystalline silicon (polysilicon) thin film transistor (TFT) technology is of great interest for active matrix organic light emitting displays (AMOLED), particularly for the latest generation of high resolution mobile phone displays [1]. The non-uniformity in the electrical characteristics of polysilicon TFTs, which is a major limitation for producing high-image-quality AMOLEDs [2], has been recently tackled by introducing increased complexity into the pixel circuits, by compensating for both threshold voltage and field effect mobility variations [3-5]. However, some compensating circuits, such as current programming methods, require very high addressing speeds for high resolution displays [4]. In addition, electrical stability is a critical requisite of the driving TFT in AMOLED applications. In order to improve circuit performance, needed for both compensating pixel circuits as well as for integrated drivers, reduction of channel length, from the typical current values of 3 µm to 1 µm or less, is being pursued. As a result, short channel effects [6] and hot carrier induced instability [7] in scaled down self-aligned (SA) polysilicon TFTs become a serious issue, and drain field engineering is mandatory. In this work we review the effects of downscaling the device geometry and, in particular, the kink effect, threshold voltage variations and hot-carrier induced instability are examined in some detail by combining experimental measurements and two-dimensional numerical simulations. In order to mitigate short channel effects and improve electrical stability, the introduction of drain field relief architectures is essential, and the advantages and disadvantages of the most popular architectures adopted to improve the device performance, including lightly doped drain (LDD) and gate overlapped LDD (GOLDD) structures, are discussed. Finally, the experimental data of the fully self-aligned GOLDD (FSA-GOLDD) polysilicon TFTs, with submicron LDD regions, are presented. We show that such advanced device architecture can provide substantial reduction of short channel effects and improve hot-carrier induced instability.

Device fabrication

Conventional self-aligned (SA) short n-channel (down to 0.4 μm) TFTs used in this work were fabricated according to a process reported in Ref. [8]. The polysilicon active layer, 40 nm thick, was crystallized by excimer laser annealing. Source and drain contacts were formed by implanting P-ions to a dose 10^{15} cm^{-2} through the oxide film, with the gate acting as a mask, and doping activation was obtained by a second pass through excimer laser. The gate oxide was deposited in a PECVD system, using SiH$_4$ and N$_2$O gas mixture, and, unless specified, the thickness was 62 nm. FSA-GOLDD polysilicon TFTs were fabricated according to the process flow reported in Ref. [9]. The initial key stage in forming FSA-GOLDD TFTs is the sputter deposition of a 1μm thick film of Al(1%Ti) as the gate metal, and its reactive ion etching to leave near-vertical side wall angles, greater than 85° for the gate pattern. The first self-aligned step is then to implant an LDD region (4 different ion doses have been adopted: $6x10^{12}$, $9x10^{12}$, $1.5x10^{13}$ and $2.5x10^{13}$ cm^{-2}) using the gate metal as the implant mask. Following this, the second key stage is to form the sidewall spacers by conformally depositing the conducting 0.5 μm thick PECVD n$^+$ amorphous silicon (a-Si) layer. Anisotropic reactive ion etching of the conformal n$^+$ a-Si layer yields the sidewall spacer features attached to the vertical sidewalls of the gate metal. The second self-aligned step uses the gate metal plus sidewall spacers as a mask for the high dose ion implantation to form source and drain contacts self-aligned to the LDD regions. One general rule of thumb from the MOSFET industry is that the width of the sidewall spacer is about 2/3rds the height of the vertical sidewall (provided the thickness of the spacer material is ≥ the gate height). Thus, with 0.5 μm thick gate metal, we can expect sub-micron self-aligned field relief regions of the order of 0.35 μm.

Short channel effects in polysilicon TFTs

Short channel effects have been well established, and widely studied, in c-Si MOSFETs [10], and they refer to a range of phenomena for which the classical long channel model no longer applies. When reducing device channel length, L, short channel effects are displayed as threshold voltage decrease with decreasing L (V_T roll-off) and with increasing source-drain voltage (V_{ds}), subthreshold slope variations and poor saturation in the output characteristics. These are a consequence of a number of inter-related phenomena, such as the size of the source and drain space charge regions becoming comparable to channel length, which reduces the amount of charge needed on the gate to invert the channel surface, and this reduces the threshold voltage with reducing channel length. Also, when the drain space charge region is comparable to channel length, such that the source and drain space charge regions start to overlap, increased drain bias can reduce the potential barrier between the source and the channel (drain-induced barrier lowering, DIBL), giving drain-bias-dependent sub-threshold currents and poor current saturation in the output characteristic. In addition, since polysilicon TFTs are fabricated on insulating substrates, floating body effects take also place [11] and are enhanced by L-reduction. In the followings, we will describe in some detail short channel effects observed in conventional self-aligned polysilicon TFTs.

Kink effect

In general, the output characteristics of polysilicon TFTs show, at high V_{ds}, an anomalous current increase, often called "kink" effect [11] in analogy with SOI devices [12]. Two-dimensional numerical simulations have shown that the kink effect is caused by impact ionization at the drain end of the channel, due to the large drain field when the device is operating in saturation. In addition, the kink effect is also enhanced by a parasitic bipolar transistor (PBT) action [13], similarly to floating body effects observed in SOI-devices. Impact ionization generated holes are injected into the floating body (base) forcing further electron injection from the source (emitter), which are then collected by the drain (collector). This added drain current augments the impact ionization, which in turn drives the floating body harder, thereby causing a regenerative action leading to a premature breakdown. Output characteristics have been measured for different L and compared at relatively low V_g (just below the threshold voltage, i.e. around 2.1 V), so that the characteristics were not affected by the parasitic resistance effect.

In Fig. 1a, typical I_d-V_{ds} characteristics are shown for different L, measured for V_g around V_T and adjusted to maintain the same low-field normalised output conductance, g_{d0}xL. As can be seen, both saturation current and kink effect appear very sensitive to channel length reduction. In particular, for short channel devices, the characteristics no longer saturate and kink effect becomes very serious. Channel length reduction also results in an increase in the output conductance, g_d, and the minimum normalised output conductance value, g_{dmin}/g_{d0}, shows an L^{-1} dependence (see Fig. 2). Output conductance increase has an adverse effect in circuit applications, as it increases in digital circuits the power dissipation and slightly degrades the switching characteristics, while, in analogue circuits, it reduces the maximum attainable gain as well as the common mode rejection ratio.

The kink effect was analyzed by using two-dimensional numerical simulations and by adopting the effective medium approximation. By using a single set of optimized parameters for the density of states (DOS) and impact ionization parameters [14], it was possible to accurately reproduce the kink effect variation with L, as can be realized by comparing experimental (Fig. 1a) and simulated (Fig. 1b) output characteristics. From the simulated output characteristics we deduced the g_{dmin}/g_{d0}, also reported in Fig. 2, and we found a very good agreement with the trend shown by the experimental data. To confirm the role of the kink effect on the drain current increase, we simulated the output characteristics by turning off the impact ionization and the corresponding curves are reported in Fig. 1b. It can be seen, by comparing the output characteristics with and without impact ionization, that the drain current at high V_{ds} is dominated by the kink effect, while DIBL has only a limited effect on the saturation drain current increase.

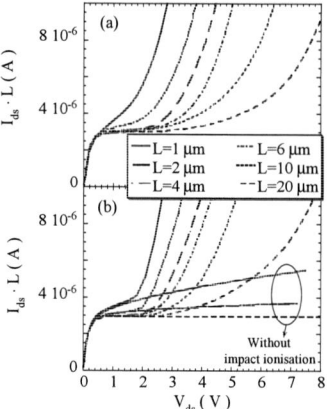

Figure 1. Experimental (a) and simulated (b) output characteristics for Vg=2.1 V (just below threshold voltage) and different channel length, L. Simulated characteristics have been calculated by including impact ionization mechanism, and for L=1, 2 and 6 µm we also show the characteristics obtained by turning off the impact ionization.

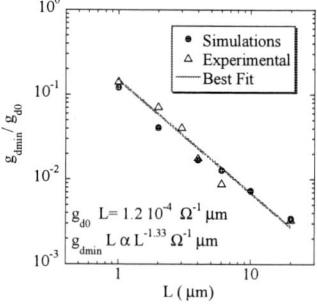

Figure 2. Experimental and simulated g_{dmin}/ g_{d0} vs channel length, L.

Threshold voltage and subthreshold slope

Threshold voltage is known to be reduced in short channel MOSFETs by decreasing channel length and increasing source-drain voltage. This effect was explained by Troutman [15] by analyzing the electrostatics of short channel MOSFETs: as the channel length is reduced, source and drain electric fields penetrate deeply into the middle of the channel, lowering the potential barrier between source and drain and determining a lower V_T with respect to the long channel case. Moreover, when a high drain voltage is applied to a short-channel device, the barrier height is lowered even more, resulting in a further V_T decrease (drain induced barrier lowering – DIBL [15]). However, in polysilicon TFTs, commonly fabricated on insulating substrates, floating body effects also represent another important factor influencing V_T. Indeed, as discussed in the previous section, the presence of high electric fields at the drain end of the channel triggers impact ionization and excess carriers are injected into the floating body giving rise to PBT effect. Therefore, in short channel polysilicon TFTs threshold voltage variations are, in general, due to a combination of mechanisms, including floating body effects, drain induced barrier lowering and field enhanced mechanisms.

In fig. 3, typical $I_d - V_g$ characteristics, measured on devices with channel widths (W) equal to 50 μm and two channel lengths at different drain bias, V_{ds}, are shown. The transfer characteristics of the shortest L show an appreciable shift of the threshold voltage, a degradation of the subthreshold slope and an increased spread in the transfer characteristics as V_{ds} increases, denoting a substantial threshold voltage variation with V_{ds}. In order to quantify the variation of threshold voltage as a function of L and V_{ds} , we defined V_T as the gate voltage at which $I_d = 10^{-7}$ A × W/L. In Fig. 4 the V_T dependence upon V_{ds} is shown for devices with channel length ranging from L=0.4 μm to L=20 μm. As can be seen, while in the long channel devices no V_T dependence upon V_{ds} is observed in the moderate and high drain bias region of the plot, in the short channel devices V_T is decreased for increasing V_{ds} , which is similar to that observed in c-Si MOSFETs. Fig. 5 shows the dependence of V_T upon the channel length for three different drain bias: it is clear that as V_{ds} is increased the V_T curves tend to spread out, denoting an increased V_T roll-off with L. This is a finding in contrast with what is observed in SOI devices, where increasing V_{ds} reduces the V_T roll-off as channel length is reduced [16].

To clarify the short channel effects we used numerical simulations, adopting as already mentioned, the effective medium approximation [11, 14]. The device characteristics were reproduced very accurately, using the set of optimized parameters reported in ref. [17] and the threshold voltage was then evaluated from simulated transfer characteristics, using the same criterion adopted for the experimental data.

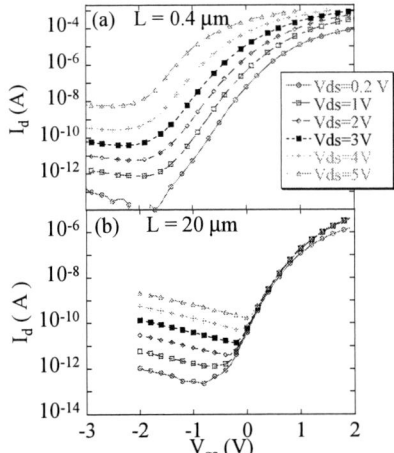

Figure 3. Experimental transfer characteristics, for a device with L = 0.4 μm (a) and L = 20 μm (b), measured at different V_{ds}.

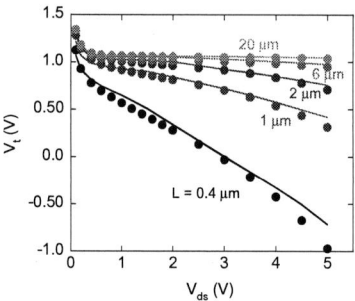

Figure 4. Experimental (symbols) and simulated (continuous lines) V_T as a function of drain bias for devices with different L. Simulations include the impact ionization model.

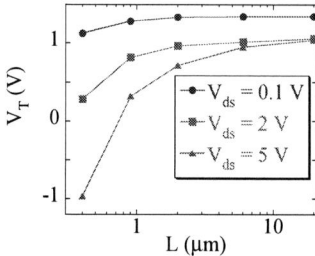

Figure 5. Threshold voltage, V_T, as a function of channel length L measured at three different drain bias.

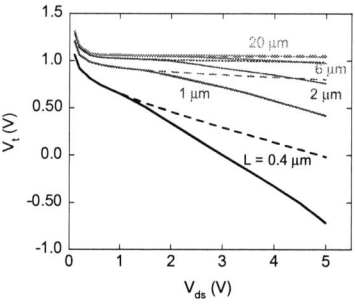

Figure 6. Simulated V_T for devices with different channel length. Simulations are performed with (continuous lines) and without (dashed lines) the impact ionization model.

The simulated V_T data nicely reproduce the dependence upon V_{ds} observed experimentally (see Fig. 4). Given the presence of impact ionization in the simulated results, the relative contributions of the direct DIBL and the PBT effects, which will also lower the source-channel barrier, can be also evaluated [17]. Figure 6 shows the simulations of V_T, with and without the impact ionization turned on, demonstrating that at low V_{ds} the DIBL was primarily responsible for the barrier lowering, but the effect of hole accumulation, via the PBT effect, reduced the barrier at higher values of V_{ds}. As would be expected from Fig. 3, both effects increased as the channel length decreased.

Hot-carrier induced instability

The presence of high electric fields at the drain junction combined with the high electron mobility leads to carrier heating. Hot-carriers can generate defects at the Si/SiO_2 interface and charge trapping in the gate oxide, inducing degradation of the device characteristics [18]. Extensive investigation of hot-carrier effects in polysilicon TFTs has shown that, similarly to c-Si MOSFETs, the device degradation is controlled by the formation of interface states and injection of charge in the gate oxide [19, 20]. By using numerical simulations, based on the interface state formation mechanism related to the sequential trapping of holes followed by electron capture (two-step model [21]), it has been demonstrated that hot-carrier induced damage is localized at the drain end of the channel at both front and back interfaces and a precise evolution of the damaged region during bias stress has been provided [19, 20].

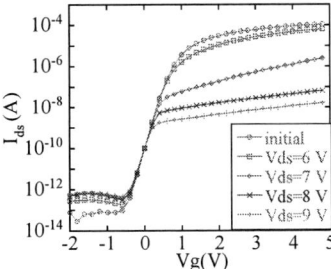

Figure 7. Transfer characteristics measured at $V_{ds}=0.1$ V during an accelerated stability test (see text) for SA polysilicon TFT with L=1.5 µm.

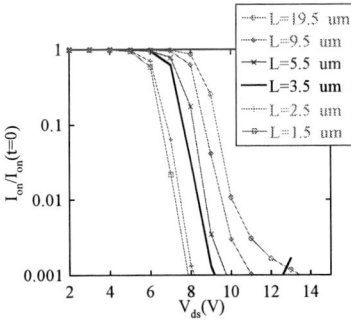

Figure 8. Relative on-current variation, measured at $V_g=5$ V and $V_{ds}=0.1$ V, during accelerated stability tests (see text) performed is SA polysilicon TFTs with different L.

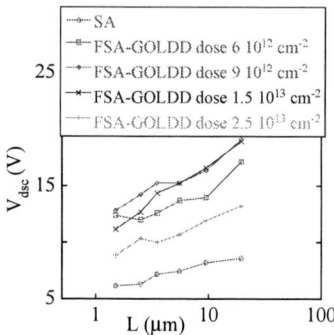

Figure 9. Critical drain bias, V_{dsc}, needed in accelerated stability test to produce 50% on-current reduction for SA and FSA-GOLDD devices with different LDD doses (as indicated) vs device channel length L.

In Fig. 7 the effects of an accelerated stability test in a SA TFT are reported: bias stress was performed for 60 seconds at $V_g=V_T$ with a given V_{ds} and the transfer characteristics were re-measured after bias stress, then the bias stress V_{ds} was incremented and the cycle repeated. As can be seen, increasing V_{ds} produces a marked degradation in the device characteristics and stronger degradation appears in short channel TFTs. In Fig. 8 the relative on-current variation vs bias stressing V_{ds} is reported for different L, showing that device degradation occurs at lower V_{ds} as L is decreased. This is more clearly evidenced in Fig. 9, where the bias-stressing V_{ds} needed to reduce to 50% of the initial on-current value (V_{dsc}) is reported for different L. From these data we can conclude that reducing L has an adverse effect on the hot-carrier induced instability, since reducing L implies higher electric fields at the drain junction, for a given V_{ds}, and, consequently, more carrier heating.

Drain field relief architectures

As shown in the previous section, short channel SA polysilicon TFTs suffer substantial electrical characteristic degradation, and drain field engineering is mandatory in view of device downscaling. Indeed, the LDD architecture has been shown to improve short channel effects [22] and electrical stability [20], although at the expenses of increased parasitic resistance [20]. Series resistance can be substantially reduced in the GOLDD structure [23], where the gate-drain overlap capacitance can be minimized by adopting a fully SA process using conductive sidewall spacers [9, 24, 25]. GOLDD structures have been demonstrated to provide excellent drain field relief with reduced series resistance [23], however, the benefits of reducing channel length could be cancelled out by series resistance and overlap capacitance from disproportionately large field relief regions. Definition of submicron LDD regions is therefore essential, and several processes have been proposed [9, 24, 25]. Originally, Hatano et al. [24] fabricated FSA-GOLDD devices by using doped polysilicon sidewall (defining LDD regions 180 nm in length) and showed the superior performance of FSA-GOLDD devices with respect to LDD or SA TFTs. The process was inspired by the well-known technology of single crystal silicon MOSFETs, namely, the use of sidewall spacers acting as sub-micron and self-aligned masks for implant and etch processes. Mishima et al. [25] also proposed a simplified method for forming self-aligned LDD regions by side etching of the Al-Nd layer in a Al-Nd/Mo terrace-structure gate electrode and using ion implantation at two different energies. However, while the Hatano et al. [24] method is a high-temperature process, requiring the deposition of polysilicon for sidewall formation, the Mishima et al. [25] approach provides poor control of the lateral extension of the LDD regions, relying upon the side etching rate of the Al-Nd layer. To solve these limitations, a process has been proposed by Glasse et al. [9] in which conductive sidewall spacers were formed by n^+ a-Si deposited by plasma enhanced CVD (PECVD), thus reducing the maximum processing temperature compared with the Hatano et al. process. In the followings we show how the introduction of FSA-GOLDD structure can effectively improve both short channel effects and electrical stability.

In Fig. 10 the output characteristics, measured in FSA-GOLDD and SA devices at the gate voltage equal to the threshold voltage ($V_g=V_T$), are shown. It is evident that the FSA-GOLDD structure effectively mitigates the kink effect, if compared to conventional SA TFTs. This is due to a reduction of the drain electric field and consequent reduction of impact ionization, and best conditions are observed for LDD doses: 6-9 10^{12} cm^{-2}.

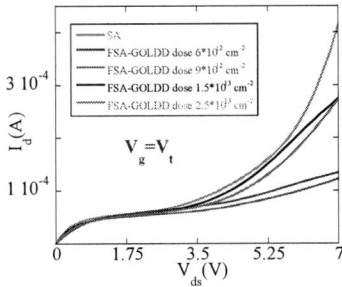

Figure 10. Output characteristics, measured at $V_g=V_T$ in L=1.5 μm SA and FSA-GOLDD TFTs with different LDD (as indicated). Gate oxide thickness was 40 nm for all devices.

To reduce the severe short channel effects present in SA polysilicon TFTs, shown in Figs. 3-5, the conventional approach of reducing the gate oxide thickness was shown to be not effective. In fact, when normalizing the V_T-variations to the oxide thickness there were no major differences, due to enhancement in the floating body effects [26]. To effectively improve the V_T-variations in short channel polysilicon TFTs, the introduction of drain field relief structures could be the only solution. Indeed, Liu at al. [22] reported a substantial reduction in the drain bias induced V_T-variations by introducing LDD regions. Indeed, the FSA-GOLDD structure appears to be quite effective, as evidenced by Fig. 11, showing the ΔV_T dependence upon V_{ds} for SA and FSA-GOLDD architectures, with different LDD doses, having defined $\Delta V_T = V_T(V_{ds}) - V_T(V_{ds}=0.1V)$, with V_T as the gate voltage at which $I_d=$ W/L 3.6 10^{-9} A. The substantial reduction in the V_T-variation at high V_{ds} in FSA-GOLDD TFTs is presumably due to reduced floating body effects.

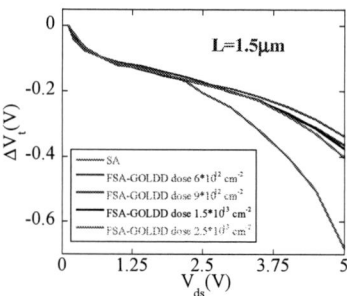

Fig. 11. Threshold voltage variation, ΔV_T, vs V_{ds} for SA and FSA-GOLDD TFTs with different LDD doses (as indicated). Gate oxide thickness was 40 nm for all devices.

To improve hot-carrier induced electrical instability in SA polysilicon TFTs, shown in Figs. 7-9, drain field relief architectures have been proved to be essential [20]. Accelerated stability tests in FSA-GOLDD devices, performed in the hot-carrier regime with similar bias conditions to the SA devices ($V_g=V_T$ and incrementing V_{ds}) are reported in Fig. 12. As can be seen, on current degradation occurs at much higher V_{ds}, if compared to SA TFTs, and the V_{dsc} values for FSA-GOLDD devices are reported in Fig. 9. Electrical stability is considerably improved in FSA-GOLDD devices, if compared to SA TFTs, and appears to be quite sensitive to LDD dose, with optimal LDD doses around 9 10^{12} cm^2.

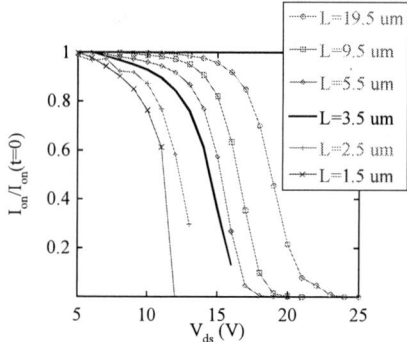

Fig. 12. Relative on-current variation, measured at $V_g=5$ V and $V_{ds}=0.1$ V, during accelerated stability tests (see text) performed is FSA-GOLDD polysilicon TFTs with 1.5 10^{13} cm^{-2} LDD dose and different L.

Conclusions

We have shown that channel length reduction, required for improving device performance, can be problematic in conventional SA polysilicon TFTs, due to short channel effects and hot-carrier induced instability. In particular, reducing channel length has a substantial impact on the kink effect, giving output conductance degradation, shallower subthreshold slope, threshold voltage roll-off and increased dependence upon V_{ds}, and enhanced hot-carrier induced instability. To improve both short channel effects and electrical stability it is essential to introduce drain field relief structures, such as LDD [20, 22] or GOLDD [23]. Indeed, the LDD architecture has been shown to improve short channel effects [22] and hot carrier induced instability [20]. As a drawback, the LDD region increases the parasitic resistance [20], and an attractive alternative is represented by the GOLDD [23] structure, which substantially reduces the parasitic resistance introduced by the LDD regions. However, for the GOLDD structure to be attractive the definition of submicron LDD regions is required. We have analyzed the electrical characteristics of FSA-GOLDD polysilicon TFTs with submicron (0.35 μm) LDD regions and different doping doses. Effective drain field relief was demonstrated by the reduction of short channel effects and the improvement of electrical stability. Best performance in terms of device stability were achieved for LDD doses in the range 9 10^{12} cm^{-2}.

Acknowledgments

The devices used in this work were fabricated at the Philips Research Laboratories, Redhill, England.

.

References

1. H.J. In, K.H. Oh, I. Lee, D.H. Ryu, S.M. Choi, K.N. Kim, H.D. Kim and O.K. Kwon, *IEEE Trans. Electron Devices*, **57**, 3012 (2010).
2. X. Guo and S. R. P. Silva, *IEEE Trans. Electron Devices*, **52**, 2379 (2005).
3. J.-C. Goh, H.-J. Chung, J. Jang, and C.-H. Han, *IEEE Electron Device Lett.*, **23**, 544 (2002).
4. Y.-H. Tai, Y.-H. Tai, B.-T. Chen, Y.-J. Kuo, C.-C. Tsai, K.-Y. Chiang, Y.-J. Wei, and H.-C. Cheng, *J. Display Technol.*, **1**, 100 (2005).
5. J.-H. Lee, W.-J. Nam, B.-K. Kim, H.-S. Choi, Y.-M. Ha, and M.-K. Han, *IEEE Electron Device Lett.*, **27**, 830 (2004).
6. G. Fortunato, M. Cuscunà, P. Gaucci, L. Maiolo, L. Mariucci, A. Pecora and A. Valletta, *ECS Trans.*, **33**, 3 (2010).
7. J.W. Lee, N. I. Lee, and C. H. Han, *IEEE Electron Devices Lett.*, **19**, 458 (1998).
8. S. D. Brotherton, S.-G. Lee, C. Glasse, J. R. Ayres, and C. Glaister, *Proc. IDW'02*, Hiroshima 4-6 December 2002, p.283 (2002).
9. C. Glasse, S.D. Brotherton, I.D. French, P. W. Green and C. Rowe, *Proc. Int. Workshop AMLCD 03*, July 2003, p. 317 (2003).
10. Y. Taur and H.T. Ning, *Fundamentals of modern VLSI devices*, Cambridge University Press, Cambridge, 2009.
11. M. Valdinoci, L. Colalongo, G. Baccarani, G. Fortunato, A. Pecora and I. Policicchio, *IEEE Trans. Electron Dev.*, **44**, 2234 (1997).
12. J. P. Colinge, *IEEE Electron Dev. Lett.*, **9**, 97 (1988).
13. J. Choi and J. Fossum, IEEE Trans. Electron Dev. **38**, 1384 (1991)
14. A. Valletta, P. Gaucci, L. Mariucci, G. Fortunato and S.D. Brotherton, *Appl. Phys. Lett.*, **85**, 3113 (2004).
15. R.R. Troutman, *IEEE Trans. Electron Devices*, **ED-26**, 461 (1979).

16. T. Tsuchiya, Y. Sato, and M. Tomizawa, *IEEE Trans. Electron Dev.*, **45**, 1116 (1998).
17. A. Valletta, P. Gaucci, L. Mariucci, A. Pecora, M. Cuscunà, L. Maiolo, and G. Fortunato, *J. Appl. Phys.*, **107**, 074505 (2010).
18. E. Takeda, C.Y. Yang, and A. Miura-Hamada, *Hot carrier effects in MOS devices*, Academic Press, San Diego, 1995.
19. L. Mariucci, G. Fortunato, R. Carluccio, A. Pecora, S. Giovannini, F. Massussi, L. Colalongo, and M. Valdinoci, *J. Appl. Phys.* **84**, 2341 (1998).
20. A. Valletta, L. Mariucci and G. Fortunato, *IEEE Trans. Electron Dev.*, vol. **53**, 43 (2006).
21. K.R. Hoffmann, C. Werner, W. Weber, and G. Dorda, *IEEE Trans. Electron Dev.* **ED-32**, 691 (1985).
22. C.T. Liu and K.H. Lee, *IEEE Electron Dev. Lett.,* **14**, 149 (1993).
23. A. Bonfiglietti, M. Cuscunà, A. Valletta, L. Mariucci, A. Pecora, G. Fortunato, S. D. Brotherton and J. R. Ayres, *IEEE Trans. Electron Dev.*, **50**, 2425 (2003).
24. M. Hatano, H. Akimoto and T. Sakai, *Proc. IEDM 97*, Dec. 1997, 523 (1997).
25. Y. Mishima and Y. Ebiko, *IEEE Trans. Electron Dev.*, **49**, 981 (2002).
26. A. Valletta, P. Gaucci, L. Mariucci, A. Pecora, M. Cuscunà, L. Maiolo, G. Fortunato and S.D. Brotherton, *Appl. Phys. Lett.*, **85**, 033507 (2009).

Self-Heating Issue of Poly-Si TFT on Glass Substrate

Tanemasa Asano and Gou Nakagawa

Graduate School of Information Science and Electrical Engineering,
Kyushu University, 744 Motooka, Nishi-ku, Fukuoka 819-0395, Japan

One of critical issues of TFT for large scale integration is temperature rise of TFT due to self-heating. In this paper, we report temperature rise of poly-Si TFT during operation. The temperature was evaluated by determining the thermal resistance from the temperature-dependent negative-drain conductance. TFTs are fabricated using a laterally-grown poly-Si film. By aligning TFT channel direction with the grain growth direction, effects of grain boundary on carrier transport becomes less significant so that direct evaluation of self-heating from drain characteristic becomes possible. SOI MOSFET is also investigated. Results indicate that the thermal resistance of TFT is 40 times as large as that of SOI MOSFET. As a consequence, temperature rise of TFT reaches to 150 K even under normal operation condition. Heat dissipation path is also investigated by determining the thermal resistance of TFTs having various dimensions. Effect of stripe channel on TFT performance and temperature rise is also discussed. Results clearly indicate that design of thermal path through the gate becomes of importance for TFTs.

Introduction

Owing to the development of crystal growth techniques for polycrystalline Si (poly-Si) thin films such as the grain positioning[1, 2, 3, 4] and lateral grain growth[5, 6, 7, 8], the current drive of poly-Si thin film transistor (TFT) has been remarkably increased. The increased current drive will promote large scale integration of TFTs to produce highly functional circuits on panel. Since total energy consumed for fabrication of TFT circuits is much less than that of transistors on single crystal Si, large scale integration technology of TFT will contribute to build up sustainable society.

On the other hand, the increased current drive of a TFT can cause unstable operation of the circuits due to temperature rise induced by *self-heating*. The thermal design becomes of important issue in developing circuit system on panel. The self-heating also significantly degrade reliability of TFT[9, 10].

To understand the self-heating effect, it is necessary to evaluate the temperature of TFTs. The self-heating has been investigated using simulation[11] or emission microscope[9, 10]. While the emission microscope is useful to evaluate the temperature distribution of the device area, it gives us the temperature not of the device active region but surface of, for example, of gate electrode. Besides the emissivity depends on the properties of the material, which causes the difficulty in determining the temperature.

On the other hand, since recent poly-Si TFT technology realizes TFTs in which carrier transport in the channel is less sensitive to grain boundaries, it becomes worth to consider evaluation of self-heating from electrical characteristics using similar manner to the one applied to metal-oxide-semiconductor field-effect transistor (MOSFET) on single crystal silicon. In this work, we investigate the temperature rise of the active region of a TFT from drain characteristic of a TFT. Using the results obtained from TFTs and SOI MOSFETs having various geometry, heat dissipation path in TFT is discussed.

Devices Under Test

TFTs investigated were fabricated on a laterally grown poly-Si film prepared by using the double-pulsed green-laser step-scan method developed by Sumitomo Heavy Industry Co. Ltd.[8, 14]. The substrate material was quartz glass whose surface was covered with SiO_2 film deposited by using plasma enhanced chemical vapor deposition. A poly-Si film deposited using low-pressure chemical vapor deposition was used as the precursor for the laser crystallization. Figure 1(a) shows scanning electron microscope showing grains of the laterally grown poly-Si film. The film composed of laterally elongated grains whose size is approximately $1\mu m$ in width and $10\mu m$ in length.

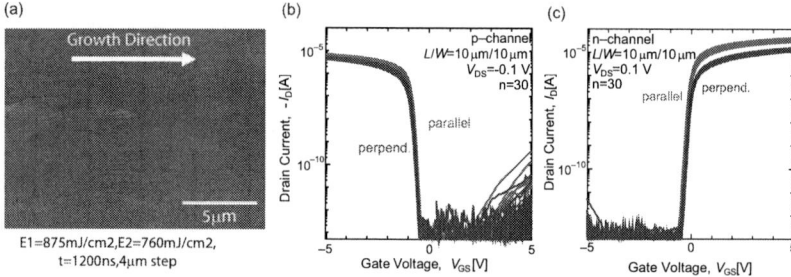

Figure 1: (a): Scanning electron micrograph of a poly-Si film prepared by the double-pulse step-scan of green laser beam. The surface of the film was chemically delineated before observation. (b) and (c): Switching characteristics of p-channel TFT and n-channel TFT, respectively, fabricated on the laterally-grown poly-Si film. Characteristics of TFTs whose channel direction is either parallel or perpendicular to the grain growth direction were plotted. Curves obtained from 30 TFTs for each type were overwritten.

TFTs whose channel is either in parallel or perpendicular to the grain-growth direction have been fabricated by using a high temperature process. The gate material of TFT was aluminum. Detail of the fabrication process is described in the literature[13]. Cross-sectional TEM analysis has shown that active region of TFT is 30 nm-thick and the gate oxide is also 30 nm-thick. Figures 1(b) and 1(c) show switching characteristics of p-channel and n-channel TFTs, respectively. Characteristics of 30 TFTs were overwritten in the figure for each type of TFT.

Field effect mobility of carriers in TFT was extracted from the transconductance. For the extraction, the effective channel length was measured from the drain characteristics of TFTs having various gate length. Figure 2 shows normal plots of n- and p-channel TFTs whose channel is either parallel or perpendicular to the grain growth direction. We can clearly see a significant anisotropy in carrier mobility in terms of channel direction with respect to the grain growth direction.

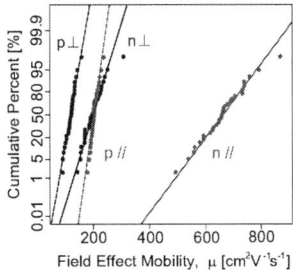

Figure 2: Normal plots showing carrier mobility extracted from switching characteristics of n- and p-channel TFTs whose channel direction were set either parallel or perpendicular to the grain growth direction, i.e, grain boundary direction.

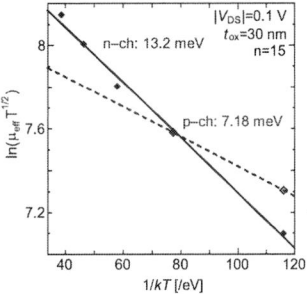

Figure 3: Arrhenius plots of temperature dependence of carrier mobility observed for TFTs whose channel direction were set perpendicular to the grain growth direction.

The carrier mobility of TFTs whose channel direction is perpendicular to the grain growth direction, i.e. grain boundary direction, is mostly determined by the transport of carriers across grain boundaries. Therefore, temperature dependence of carrier mobility of these TFTs gives us the value of potential barrier generated at grain boundary. Figure 3 shows Arrhenius plots of carrier mobility obtained from TFTs whose channel is set perpendicular to the direction of grain boundaries. The plots gives us about 13 mV and 7 mV for n- and p-channel, respectively, while the value is dependent on gate bias. The thermionic emission model taking these values into account reasonably explain the observed carrier mobility.

On the other hand, carrier transport in TFTs whose channel direction is parallel to the grain boundary direction are almost free from the presence of grain boundaries and is determined by phonon scattering unless TFT is operated at very low temperatures. Therefore, temperature rise due to self-heating can be measured by using such high mobility TFTs.

Model to Determine the Temperature Rise

The temperature rise of TFT is evaluated by using the model originally developed by Jomaah et al [12] to investigate the self-heating of SOI MOSFET. In principle, the model uses the drain conductance appeared in the saturation region of MOSFET operation due to self-heating. The drain conductance in the saturation region g_{Dsat} becomes negative when self-heating is significant mostly due to reduction of carrier mobility with temperature rise.

The drain conductance g_{Dsat} in the saturation region is described by

$$g_{\mathrm{Dsat}} = \frac{\mathrm{d}I_{\mathrm{D}}}{\mathrm{d}V_{\mathrm{DS}}} = \frac{\mathrm{d}I_{\mathrm{D}}}{\mathrm{d}T}\frac{\mathrm{d}T}{\mathrm{d}V_{\mathrm{DS}}}, \tag{1}$$

where T is temperature of the device under test, I_{D} is drain current, V_{DS} is drain voltage. Temperature change $\mathrm{d}T$ can be expressed by using a thermal resistance R_{th} from the channel to a heat sink

$$\mathrm{d}T = R_{\mathrm{th}}\mathrm{d}\left(I_{\mathrm{D}}V_{\mathrm{DS}}\right). \tag{2}$$

Since, in the saturation region, the drain current does not depend on V_{DS} but is constant at I_{Dsat},

$$\mathrm{d}T = R_{\mathrm{th}}I_{\mathrm{Dsat}}\mathrm{d}V_{\mathrm{DS}}. \tag{3}$$

Equation (1) therefore becomes

$$g_{\mathrm{Dsat}} = \frac{\mathrm{d}I_{\mathrm{D}}}{\mathrm{d}T}R_{\mathrm{th}}I_{\mathrm{Dsat}}. \tag{4}$$

If we know the temperature dependence of drain current, $\frac{\mathrm{d}I_D}{\mathrm{d}T}$, by changing the substrate temperature, we can determine R_{th} by using the above equation from g_{Dsat} measured at a certain substrate temperature.

Once R_{th} is determined, the temperature rise ΔT due to self-heating can be evaluated by

$$\Delta T = R_{th}I_{D}V_{DS}. \tag{5}$$

The R_{th} is in general dependent on temperature and is expressed as

$$R_{\mathrm{th}}^{*} = \frac{R_{\mathrm{th}}}{1 + I_{\mathrm{D}}V_{\mathrm{DS}}\mathrm{d}R_{\mathrm{th}}/\mathrm{d}T} \tag{6}$$

When the temperature dependence $\mathrm{d}R_{\mathrm{th}}/\mathrm{d}T$ is small and the relation $1 \gg I_{\mathrm{D}}V_{\mathrm{DS}}\mathrm{d}R_{\mathrm{th}}/\mathrm{d}T$ holds, we can assume $R_{\mathrm{th}} \simeq R_{\mathrm{th}}^{*}$. We will find this assumption is quite reasonable in determining the temperature rise of TFTs and SOI MOSFET.

Results and Discussion

The characteristics of TFTs were measured to determine dI_D/dT of equation (4) in the temperature range from 100K to 300K at the interval of 50K using a probing system equipped with a liquid He cryo system. SOI MOSFETs of partially depleted type were also fabricated on a separation by implanted oxygen (SIMOX) wafer having the buried oxide thickness of about 160 nm. The gate oxide thickness of SOI MOSFET was 10 nm. The gate material was poly-Si.

To investigate effect of TFT design on self-heating, we have fabricated and tested two layout designs, single gate TFT and TFT composed of multiple stripes[15]. The layout examples are shown in Fig. 4 where the single gate is of $W/L = 10\mu m/10\mu m$ and the striped gate is of $L = 10\mu m$ and $W = 2\mu m \times 5$. The single gate TFT was designed to have the body tie structure to minimize the floating body effect.

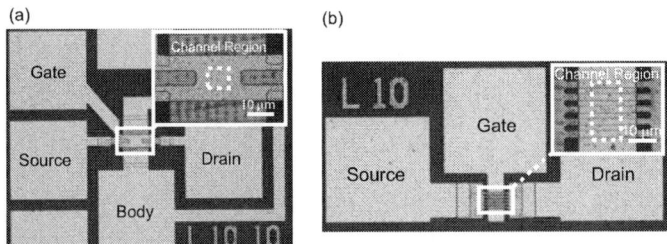

Figure 4: Photographs showing layouts of (a) single gate TFT of $W/L = 10\mu m/10\mu m$ and (b) multiple stripe-gate TFT of $L = 10\mu m$ and $W = 2\mu m \times 5$.

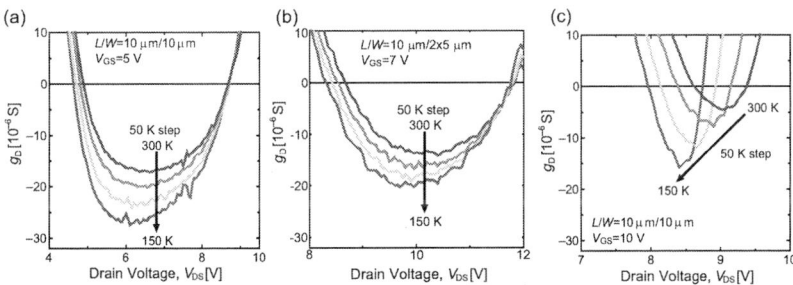

Figure 5: Drain conductance g_D observed for (a) single gate TFT, (b) multiple stripe-gate TFT, and (c) single gate SOI-MOSFET. g_D was extracted from drain characteristics measured at temperatures from 300 K to 150 K.

Thermal Resistance R_{th} and Self-heating

Figure 5 shows change in drain conductance with drain voltage for single gate TFT, stripe-gate TFT and single gate SOI-MOSFET operated at temperatures between 150 K and 300 K. The appearance of the negative drain conductance is clearly observed. It is noteworthy that TFTs whose channel direction is perpendicular to the grain growth direction did not show negative drain conductance. The appearance of the negative drain conductance is pronounced in TFT compared with SOI MOSFET, suggesting that the temperature rise is much higher in TFT than in SOI MOSFET.

Detail of the process to determine the thermal resistance R_{th} using dI_D/dT and g_D values has been described in the previous publication[16]. Resultant R_{th}'s are shown in Fig. 6(a) as a function of device operating temperature. We find that the thermal resistance R_{th} of single gate TFT is 40 times as large as that of SOI MOSFET. However, R_{th} can be significantly reduced by employing multiple stripe-gate structure.

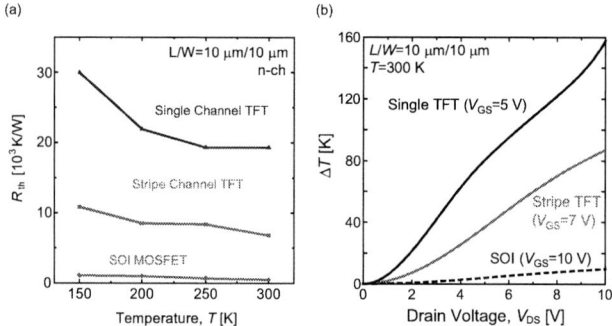

Figure 6: (a) Thermal resistances R_{th}'s obtained for single gate TFT, multiple stripe-gate TFT, and single gate SOI-MOSFET as a function of temperature. Effective transistor size was $W/L = 10\mu m/10\mu m$ for all transistors. (b) Temperature rise ΔT for three devices.

From the thermal resistance R_{th} we can evaluate the temperature rise ΔT using equation (5). Results are shown in Fig. 6(b). It is noteworthy that the $I_D V_{DS} dR_{th}/dT$ values were 0.005 and -0.075 for single gate TFT and single gate SOI at $V_{DS} = 10$ V and $T = 300$ K. These values are much smaller than unity, indicating that the assumption $R_{th} = R_{th}^*$ is quite reasonable. At $V_{DS} = 10$V, the temperature rise of TFT is about 160 K, while that of SOI is only about 10 K. Multiple stripe-gate structure can reduce the temperature rise to about 80 K .

Discussion on Thermal Path

In order to investigate the thermal path of TFT and SOI MOSFET, change in heat resistance with channel length and channel width has been investigated[16]. Results indicated that the heat resistance of SOI MOSFET was inversely proportional to the channel area while the heat resistance of TFT was not in proportion to the inverse of the channel area.

From these results, we can model the heat path for TFT and SOI MOSFET as the schematic illustration shown in Fig. 7. In the case of SOI MOSFET, significant heat dissipation takes place through the buried oxide to the Si substrate. On the other hand, in the case of TFT, there is little heat dissipation through the substrate because of the very low thermal conductivity of the glass substrate. Heat dissipation of TFT mostly takes place through the gate oxide and gate electrodes. Therefore, multiple stripe-gate structure provides less thermal resistance than the single gate structure. Heat is also dissipated through source and drain but this portion is not significant because Si is thin.

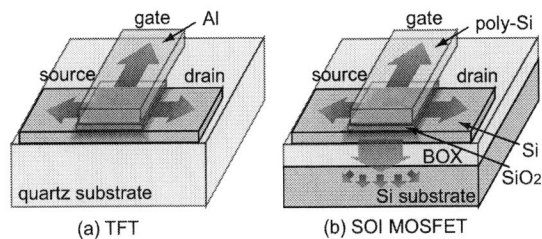

Figure 7: Heat path model for (a) TFT and (b) SOI MOSFET.

In order to examine the validity of the heat path model, we have estimated the thermal resistance of TFT and SOI by assuming the simple model which employs

$$R_{\text{th}} = \Sigma_{m=1}^{n} \frac{1}{\lambda_m} \frac{l_m}{S_m}$$

where λ_m, l_m, and S_m are thermal conductivity, length, and cross-section of the material m which are in touch with the active region of the TFT and MOSFET. We used the thermal conductivity of each material as follows; $\lambda_{\text{Si}} = 168$ W/mK, $\lambda_{\text{SiO}_2} = 1.4$ W/mK, $\lambda_{\text{Al}} = 236$W/mK for Si, SiO$_2$, and aluminum, respectively. The calculated thermal resistances are listed in Table 1 together with experimentally observed values. We find that the calculated values are in reasonable agreement with the experimentally determined values in spite of the simplicity of the model.

Table 1: Calculated and experimentally determined values of heat resistance.

	Single Gate TFT	Stripe Gate TFT	Single Gate SOI-MOS
Calculated	26,000	7,400	720
Experimental	19,300	6,980	485

Conclusion

We have investigated temperature rise due to self-heating and heat flow path of poly-Si TFTs fabricated using a laterally grown poly-Si film on quartz glass substrate. Elimination of potential barrier for carrier flow in the channel of TFT enabled us the direct evaluation of self-heating from the negative drain conductance. It has been found that the heat resistance of TFT is about 40 times as large as that of SOI MOSFET having 160 nm-thick buried oxide due to the very poor thermal conductance of the substrate. In case of TFT, heat flow through the gate electrode is significant and important. Multiple stripe-gate has been proposed and verified to increase the heat flow and to decrease the temperature rise.

Acknowledgment

The author are grateful to Kazunori Watanabe and Koji Akiyama for their help in experimental. They are also thankful to Takayuki Takao for his excellent technical assistance. This work is supported in part by the Grant-in-Aid for Scientific Research (No. 21246061) from Japan Society of Promotion of Science.

References

[1] T. Endo, Y. Taniguchi, T. Katou, S. Shimoto, T. Ohno, K. Azuma, and M. Matsumura: Jpn. J. Appl. Phys. **47** (2008) 1862.

[2] K. Makihira and T. Asano: Appl. Phys. Lett.**76** (2000) 3774.

[3] K. Makihira, M. Yoshii and T. Asano : Jpn. J. Appl. Phys. **42** (2003) 1983.

[4] G. Nakagawa and T. Asano: Jpn. J. Appl. Phys. **45** (2006) L1293.

[5] J. Im: Appl. Phys. Lett. **64** (1994) 2303.

[6] M. Hatano, H. Akimoto, T. Sakai and H. Ishida: *Dig. Tech. Papers, Active-Matrix Liquid Crystal Displays*,Tokyo, 1997, p. 95.

[7] A. Hara, F. Takeuchi, M. Takei, K. Suga, K. Yoshino, M. Chida, Y. Sano, and N. Sasaki, *Dig. Tech. Papers, Active-Matrix Liquid Crystal Displays*, Tokyo, 2002, p. 227.

[8] S. Sakuragi, T. Kudo, Y. Yamazaki, and T. Asano: *Proc. Int. Display Workshops, 2005*, AMD7-4.

[9] S. Inoue, M. Kimura and T. Shimoda: Jpn. J. Appl. Phys. **42** (2003) 1168.

[10] S. Hashimoto, Y. Uraoka, T. Fuyuki and Y. Morita: Jpn. J. Appl. Phys. **45** (2006) 7.

[11] T. Shimatani, T. Matsumoto, T. Hashimoto, N. Kato, S. Yamada and M. Koyanagi: Jpn. J. Appl. Phys. **33** (1994) 619.

[12] J. Jomaah, G. Ghibaudo and F. Balestra: Solid State Electron. **38** (1995) 615.

[13] T. Asano, K. Watanabe, M. Esaki, G. Nakagawa, S. Sakuragi, T. Kudo, K. Yamasaki: *Proc. 2nd International TFT Conference, Kitakyushu, 2006*, p. 154.

[14] K. Watanabe and T. Asano: *Proc. 3rd International TFT Conference, Seoul, 2006*, p. 77.

[15] K. Akiyama, K. Watanabe, and T. Asano: Jpn. J. Appl. Phys. **48** (2009) 03B014.

[16] K. Watanabe and T, Asano: Jpn. J. Appl. Phys. **48** (2009) 03B005.

Bridged-grain (BG) Polycrystalline Silicon Thin Film Transistors (TFTs)

H. S. Kwok, M. Wong, S. Y. Zhao and W. Zhou

Department of Electronic and Computer Engineering, Hong Kong University of Science and Technology, Clear Water Bay, Kowloon, Hong Kong

A new technique bridged-grain (BG) was introduced. Using this BG excimer laser annealing (ELA) poly-Si as an active layer, the grain boundary effects can be minimized. Important electrical properties such as sub-threshold swing, threshold voltage, maximum field effect mobility, leakage current, on-off ratio and device uniformity across the substrate can all be improved using the present technique. The improvement can be achieved at low cost, thus making inexpensive, high performance LTPS TFT a reality.

Introduction

Among existing display technologies, active-matrix organic light emitting diodes (AMOLEDs) is regarded as the best candidate for an 'ultimate display' due to their superior characteristics, such as self-emission, fast motion picture response, vivid colors, wide view angle, high contrast, and super-slim [1]. P-channel low temperature poly silicon (LTPS) thin film transistors (TFTs) have attracted more and more attention due to their superb current-driving capability and electrical reliability. It is well known that the LTPS has the potential for peripheral circuit integration to realize system on panel (SOP) due to the higher mobility than a-Si TFTs. To reduce the manufacture cost and mask number, a lot of PMOS circuits for display have been proposed, including shift registers, level shifters, demultiplexers, and DC-DC converter [2]. Thus low cost, high performance and reliable LTPS processing technologies are greatly required [3]. Considering the trade-off between performance and economic issue, three kinds of LTPS technology survived out of multifarious approaches, each with its own advantages and disadvantages. They are solid phase crystallization (SPC) [4], excimer laser annealing (ELA) [5] and metal induced crystallization (MIC) [6].

Among these methods, ELA has been the most widely used due to its excellent crystallization quality, fast crystallization speed, and high mobility. In 2004, 2 inch System on Glass (SOG) QVGA (240×320) TFT-LCD with integrated 6-bit source driver circuit was fabricated employing the high-performance Two-Shot Sequential Lateral Solidification (TS-SLS) TFTs [7]. In 2008 SID, Samsung demonstrated 14 inch WXGA AMOLED with SLS base TFT backplane [8]. They already had some commercial products in mobile display, such as i9000 Galaxy S with a 4 inch 480×800 LTPS-AMOLED panel.

Even ELA is already employed in mass-production, the ELA poly-Si TFT still suffers from the non-uniformity of TFT current output due to its differences in threshold voltage, mobility and sub-threshold swing from different degree of crystallization of TFTs. The non-uniform brightness from pixel to pixel was observed. In addition, one of the important problems of ELA-TFTs is large off-state current. An important approach for reducing leakage current is to decrease the electric field in the drain depletion region. Normally it was achieved by replacing poly-Si TFTs with a self-aligned structure by those with a lighted doped drain (LDD) structure[9].

In this paper, we proposed a new method to improve the properties of ELA TFTs. Using bridged-grain (BG) ELA poly-Si as an active layer, the grain boundary effects can be significantly reduced. Important properties such as threshold voltage, sub-threshold slope, device mobility, on-off ratio and device uniformity across the substrate can all be improved using the present technique. The improvement can be achieved at low cost, thus making inexpensive, high performance LTPS TFT a reality.

Fabrication of ELA-BG poly-Si TFT

These BG-ELA poly-Si films were patterned into active islands by dry etching with AME8110 reactive ion etcher. After dry etching, photoresist was removed by O_2 plasma. 100nm low temperature oxide (LTO) was subsequently deposited by low pressure chemical vapor deposition (LPCVD) at 425°C as the gate insulator after the native oxide was removed by 1% HF. Following the deposition of 200nm Titanium which was patterned into gate electrodes, boron and phosphor at a dose of 4×10^{15}/cm^2 was implanted into the S/D for p-channel TFT and n-channel TFT, respectively. A 500nm LTO isolation layer was deposited. Contact holes were opened before 700nm aluminum-1%Si was sputtered and patterned as electrode contacts. Contact sintering was then performed in forming gas annealing (FGA) at 420°C for 30mins.

Device characterization and discussion

Electrical characteristics of both p-channel and n-channel TFTs built on 45nm ELA and ELA-BG poly-Si were measured with HP4156 semiconductor parameter analyzer. Firstly, transfer characteristic curves for TFTs of which W/L was 30μm/10μm were tested. W and L denote the channel width and gate length of the transistor respectively, as shown in Figure 1. The threshold voltage (V_{th}) defined as the V_g required to induce an I_d of W/L×10^{-7} A at $|V_{ds}|$=5V.

The field-effect mobility (μ_{FE}) of normal ELA TFT at low drain voltage is given by

$$\mu_{FE} = \frac{Lg_m}{WC_{ox}V_{ds}} \qquad [1]$$

Where g_m is the transconductance, C_{ox} is the gate insulator capacitance per unit area, V_{ds} is the voltage between drain and source. The reported field-effect mobility is the maximum value measured.

Figure 2 shows the transfer characteristics of p-channel ELA and BG-ELA TFTs when V_{ds}=-0.1V and V_{ds}=-5V. Figure 3 shows the transfer characteristics of n-channel ELA and BG-ELA TFTs when Vds=0.1V and Vds=5V. The important electrical parameters of BG-ELA and ELA TFTs are listed in Table 1, including both p- and n-channel TFTs.

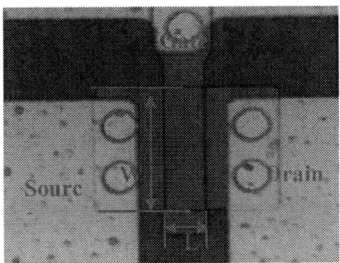

Figure 1. Microscope photo of the TFT.

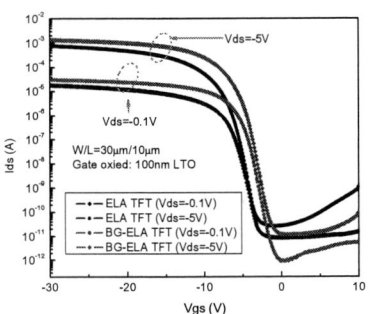

Figure 2. The transconductive curves of p-channel BG-ELA and ELA TFTs, when Vds=-0.1V and Vds=-5V.

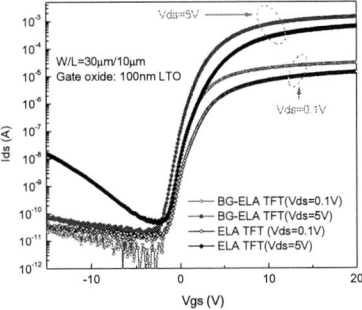

Figure 3. The transconductive curves of n-channel BG-ELA and ELA TFTs, when Vds=0.1V and Vds=5V.

The effective channel length of BG-TFT is changed due to the BG structure. To better investigate the BG effects, the IV curves of BG-ELA TFTs are need to be processed according to the channel length modulation by BG structure. Here, we assume that, the field effect mobility of ELA-poly-Si at "on" state remains unchanged whether there is BG or not, when the TFT are working at "on" state. When Vds=-5V, the "on" state current of normal p-channel ELA TFT with BG and without BG is 1.35×10^{-3} A and 7.56×10^{-4} A respctviely. The current ratio between p-channel TFTs with BG and without BG is 1.78. The designed channel width and length for TFTs are 30μm and 10μm, respectively. So, the effective channel length of ELA TFTs with BG is the length of gate to be devided by the factor of 1.78. As a result, the effective channel length of BG-ELA TFTs for p-channel and n-channel are 5.62μm and 4.59μm, respectively. As shown in table 1, the maximum field effect mobility of TFTs with BG structure are increased by 30%.

Table 1. Key device parameters of BG-ELA and normal ELA TFTs, including both n-channel and p-channel.

	P-Channel		N-Channel	
	ELA	**BG+ELA**	**ELA**	**BG+ELA**
μ_{FE}(cm^2/Vs)	101	139	105	134
V_{th} (V)	-6	-4	1.3	0.4
S (V/decade)	0.72	0.6	0.734	0.64
I_{off} (pA/µm) Minimum	0.9	0.39	1.62	0.437
GIDL (pA/µm) $\mid V_{ds} \mid$ =5V $\mid V_{gs} \mid$ =10V	25	3.23	52.67	1.25
I_{on}/I_{off} ($\times 10^7$) $\mid V_{ds} \mid$ =5V	1.78	11.25	1.37	11.1

The subthreshold swing (S) is 0.6V/dec and 0.72V/dec for p-channel ELA TFTs with BG and without BG, respectively. Also, with BG, the absolute value of threshold voltage (V_{th}) of p-channel TFT is reduced by 2V to 4V. The subthreshold swing (S) is 0.64V/dec and 0.73V/dec for n-channel ELA TFTs with BG and without BG, respectively. Also, with BG, the threshold voltage (V_{th}) of n-channel TFT is reduced by 0.9V to 0.4V. Normally, the V_{th} and S are strongly influenced by the deep trap state, associated with danling bonds [10]. In other words, we believe that the boundary height could be lowered by the BG structure, meanwhile, some of the trap states and dangling bounds could be

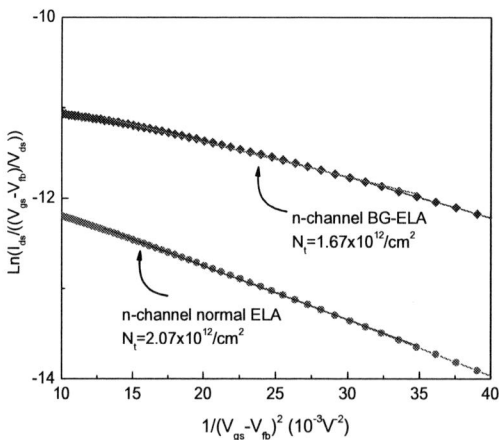

Figure 4. Plots of $\ln(I_{ds}/((V_{gs}-V_{fb})/V_{ds})$ *vs* $1/(V_{gs}-V_{fb})^2$ for the n-type normal ELA and BG ELA TFTs

filled or terminated by the BG structure. Extracted from Figure 4 using a model proposed by R. E. Proano [10], the respective trap state densities for the normal ELA and BG ELA TFTs are $2.07 \times 10^{12}/\text{cm}^2$ and $1.67 \times 10^{12}/\text{cm}^2$.

As shown in Figure 2 and Figure 3, the leakage current for both n-channel and p-channel TFTs was greatly reduced by applying BG structures to the active channel. At high V_{gs} and high V_{ds}, the leakage current is dramatically reduced by a more than one order. As a result, the on/off ratio is also greatly improved, as listed in Table.1. When V_{gs} is smaller than V_{th}, the ELA poly-Si without implantation is intrinsic and the doped region presents N+ or P+ polarity in case of n-channel or p-channel TFTs, which means the active channel for TFTs with BG structure becomes a series of i-N+ or i-P+ shallow junctions in this case. That is why the GIDL and minimum current are both greatly decreased. From the above comparison, we can see most of the TFT parameters were dramatically improved with BG apllication.

Figure 5 (a) and (b) show the p-channel TFT performance variation, V_{th} and GIDL, for normal ELA TFTs and BG-ELA TFTs. The data was measured for 98 TFTs uniformly distributed over 4 inch glass wafers. It is clearly that the TFTs with BG structrue show much lower GIDL than normal ELA TFTs, so as to the uniformity. Meanwhile, the BG-ELA TFTs also present smaller V_{th} variation than normal ELA TFTs, as well as the absolute V_{th} value.

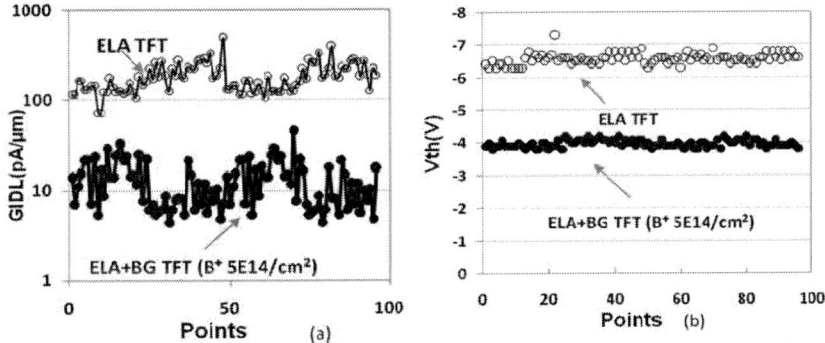

Figure 5. TFT performance variation for uniformly distributed 100 TFTs; (a) GIDL and (b) V_{th}.

Conclusion

A bridged-grain (BG) technology was proposed to apply to commercial ELA TFTs. The ELA TFTs with BG structure shows smaller threshold voltage, steeper subthreshold slope, smaller leakage current, higher on-off ratio, and better device uniformity across the substrate. The improvement can be achieved at low cost, thus making inexpensive, high performance LTPS TFT a reality.

Acknowledgments

This work was supported by Hong Kong Government Research Grants Council and grant number 614807

References

1. Lee, S.T., et al., *53.1: Invited Paper: LITI (Laser Induced Thermal Imaging) Technology for High-Resolution and Large-Sized AMOLED.* SID Symposium Digest of Technical Papers, 2007. **38**(1): p. 1588-1591.
2. Yeh, S.-H., et al., *P-173L: Late-News Poster: A 2.2-inch QVGA System-on-Glass LCD Using P-Type Low Temperature Poly-Silicon Thin Film Transistors.* SID Symposium Digest of Technical Papers, 2005. **36**(1): p. 352-355.
3. Kim, H.D., et al., *22.1: Invited Paper: Technological Challenges for Large-Size AMOLED Display.* SID Symposium Digest of Technical Papers, 2008. **39**(1): p. 291-294.
4. Kim, M., et al., *Effects of high pressure annealing on the characteristics of solid phase crystallization poly-Si thin-film transistors.* Journal of Applied Physics, 2008. **103**(4): p. 044508-044508-5.
5. Kubo, N., et al., *Characteristics of polycrystalline-Si thin film transistors fabricated by excimer laser annealing method.* Electron Devices, IEEE Transactions on, 1994. **41**(10): p. 1876-1879.
6. Meng, Z., *Metal-Induced Unilaterally Crystallized Polycrystalline Silicon Thin-Film Transistor Technology and Application to Flat-Panel Displays*, in *Electrical and Electronic Engineering*. 2002, The Hong Kong University of Science and Technology. p. 2.
7. Kim, C.W., et al., *21.4: Development of SLS-Based System on Glass Display.* SID Symposium Digest of Technical Papers, 2004. **35**(1): p. 868-871.
8. Choi, J.B., et al., *9.2: Sequential Lateral Solidification (SLS) Process for Large Area AMOLED.* SID Symposium Digest of Technical Papers, 2008. **39**(1): p. 97-100.
9. Orouji, A.A. and M.J. Kumar, *Leakage current reduction techniques in poly-Si TFTs for active matrix liquid crystal displays: a comprehensive study.* Device and Materials Reliability, IEEE Transactions on, 2006. **6**(2): p. 315-325.
10. Proano, R.E., R.S. Misage, and D.G. Ast, *Development and electrical properties of undoped polycrystalline silicon thin-film transistors.* Electron Devices, IEEE Transactions on, 1989. **36**(9): p. 1915-1922.

Vertical Channel Thin Film Transistor: Improvement approach similar to multigate monolithic CMOS Technology

Olivier Bonnaud, Peng Zhang, Emmanuel Jacques, Régis Rogel

Groupe Microélectronique IETR UMR 6164, University of Rennes 1 - Campus de Beaulieu, 35042 RENNES Cedex France,

The silicon based thin film transistors (TFTs) technology evolution is governed by electrical performances that are needed for many applications. Previous papers highlighted the influence of the process, the quality and the nature of the channel material on their electrical properties. TFTs had as well channel parallel to the substrate surface as quasi-vertical channels. This last choice allows an increase of the equivalent current density flowing in the circuit thanks to a shorter channel length and a higher channel width, for the same lithographical node. However, the drain-source leakage current was measured proportional to the source-drain face-to-face area, while I_{on} is proportional to channel width. This behavior was also observed in vertical monolithic structures. To minimize this effect two main ways are proposed: decreasing the scale, or including a physical barrier between source and drain. We will see that these approaches are really similar to ULSI technology ones.

Introduction

The silicon based thin film transistors (TFTs) technological process evolution is governed by electrical performance improvements that are needed for the numerous fields of large area electronic applications. The main objective in agreement with these applications is to fabricate TFTs at a temperature low enough to be compatible with the substrates, such as glass substrates, integrated circuit processed substrates (above IC devices), and flexible substrates, that implies several approaches with good enough electrical properties. Previous papers highlighted on one hand the influence of the process and on the other hand the quality and the nature of the channel material on the electrical properties of the TFTs, mainly equivalent mobility of carriers in the channel, I_{on}/I_{off} drain current ratio, and subthreshold slope (1). The previous studies involved planar process that means channel parallel to the substrate surface. Another way passes by a modification of the spatial geometry, more especially by the creation of vertical channels. This allows an increase of the equivalent current density flowing in the circuit thanks to two major modifications, the channel length that can be much shorter and the channel width that can be much higher, for the same design rules, i.e. for the same lithographical node. The approach is thus similar to the recent evolution of the monolithic silicon technology that proposed FinFET (2-4), double gate and gate-all around MOSFETs (5-6), and vertical channel FET structures (7-12), that lead to significant expected improvements and open some new domains of circuit applications (3).

In the field of thin film technologies several approaches were made to move to multi-gate devices. The first published works in a journal on this topic concerned the organic electronics (13). Its main advantage was to fabricate a transistor with double channels and a very thin channel region that can be fully depleted. A structure on insulator (SOI) fully depleted, allowed some interesting improvements, mainly the breakdown voltage thanks to a lightly doped channel region and higher carrier mobility in the channel. In the case of thin film technology, an undoped channel region combined with a vertical stacking of drain, channel, and source zones, may lead to similar improvements of the global electrical features.

Even though the potential interest of such a structure appears very attractive, the leakage current between drain and source was observed as directly dependent of the face-to-face common area between source and drain, while the on current, I_{on}, is proportional to the channel width. This behavior was also observed in vertical monolithic structures such as multigate MOSFET and gate-around MOSFET. To minimize this leakage current that affects drastically the I_{on}/I_{off} ratio, a first way consists to decrease the lateral geometry, in this case the distance between the two vertical channels. This means an improvement of the alignment techniques that is difficult to insure due to the relatively high thicknesses of the films. In other words, even the lithographical node is better the accuracy of the pattern size is affected by the global thickness of the stacking. The second way consists to incorporate a thin film between source and drain acting as a current barrier, the channel region being deposited just on the quasi vertical sidewalls. The recent studies were oriented on this solution.

After the presentation of the interest of such vertical channel ways on both ULSI and thin film technologies, a review on the different solutions already proposed is given. On the base of new vertical multi-tooth thin film transistor results (14) presented at the IXth Thin Film Transistor Technology Conference 2008, some specific behaviors are presented and discussed. A special attention is paid on the technological steps involved in the vertical (or quasi-vertical) channel fabrication, in order to decrease the leakage current. The proposed approaches are similar to the ULSI ones, with the same final goal: improve the total current flowing through the same area of substrate, decrease, the leakage current in order to maintain the I_{on}/I_{off} ratio, and decrease the parasitic capacitances.

Architecture evolution towards vertical channel transistor

In the early silicon technologies, planar processes were first used in the development of integrated circuit. Since the beginning of the sixties, the technologies have been evolving following the Moore's law (15) with an exponential decreasing of the size of the elementary transistor associated to an exponential increasing of the number of components in an integrated circuit. This evolution was possible thanks to a permanent decrease of the lateral geometry associated to self aligned process steps that minimized the number of photolithography steps. By decreasing the wavelength of the insolating equipment, the diffraction limit was pushed down and the pattern were thus smaller. In order to go further, some hybrid equipments were set-up; they mainly involve electron beam based insolating, the associated wavelength being much smaller in this case. However, the limitations mainly due to the cost of equipment became a drawback and some evolution of the architecture was able to maintain this terrific Moore's law evolution. The new technologies are presently moving to three-dimensional geometries that should allow continuing this fabulous evolution.

As mentioned above, traditional MOS devices deploy transistors onto the surface of the silicon in a planar or horizontal fashion. In recent years new device architectures have emerged, featuring vertical transistors that utilize multiple sides of the silicon. The first structures moving to 3D- architecture were designed with lateral walls for the gate contacts. Vertical transistors having more than one gate to control the device are appealing in part because they reduce leakage currents and they provide higher drive current functions. But vertical transistors present fundamental design and manufacturing challenges related to mechanically stability, sub-lithographic feature sizes and patterning over tall topographies. Several solutions were proposed during the last ten years. A new family of MOSFETs with a fin for the channel region was created; the corresponding transistor, called FinFET is schematized on figure 1 (2-4). Source and drain are laterally located and the channels are generated on lateral walls; this leads to a higher equivalent width of the channel for the same area of the transistor. Other geometries were proposed such as omega gate, gate around (5) or inverted T-shape (3-4). This last one, proposed by a Freescale team (3) is shown figure 2. The channel section has a T-shape and the channel is lightly doped that leads to a fully depleted transistor. The advantage of the fully depleted devices, mainly developed on SOI substrates, comes from the lowering of the drain induced barrier leakage current (DIBL) and of the short channel effects (SCE). They have also a higher breakdown voltage and a higher driven current for the same substrate area.

For all these structures, the drain-source current remains parallel to the substrate.

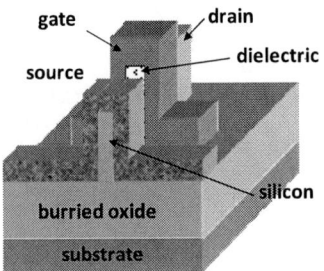

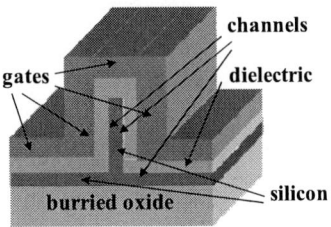

Figure 1.   Schematic   perspective representation   of   a   FinFET.   The channels are no longer parallel to the surface of the substrate. The channel width corresponds to the height of the both vertical walls of the fin (2).

Figure 2.   Inverted T-Fin Field Effect Transistor. (3). The channels are both parallel and perpendicular to the surface. The conduction remains parallel.

The true vertical channel transistor means a drain-source current flowing perpendicularly to the substrate, the drain on the top and the source on the bottom, or the contrary. The main advantages of such a structure come from i) the control of the channel length by the thickness of the channel region that can be very well controlled thanks to the growth control of the films (by epitaxy or CVD deposition techniques for example), ii) the width of the channel that is the addition of the both lateral gate widths.

In the early 2000's several solutions were proposed. The principle was published by Orlowski et al. (16) and analyzed by Moers (7). A schematic cross-section of the vertical MOSFET is shown in figure 3. The distance between the both gates can be the design rule parameter of the deca-nanometric technology that means much smaller than the tenth of micrometer. Potentially, for ultra large scale integration (ULSI), this architecture is very promising as mentioned by the authors.

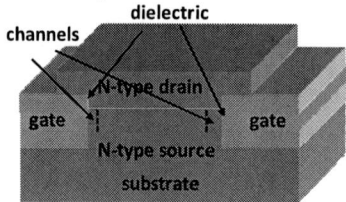

Figure 3: Principle of a vertical MOSFET. The both lateral channels are vertical and the drain-source current flows perpendicularly to the substrate surface (*after* Orlowski *et al*. (16)).

However, if this structure is not yet fabricated and not yet described in the literature in monolithic integrated circuit technologies, the architecture is already available for thin film technologies. The first published works in a journal concerned the organic electronics (13). The stacking of nine films, four of them being organic base compounds (pentacene and diphenil-benzidine), allowed getting electrical characteristics with the I_{ON}/I_{OFF} current ratio about 10^3, and the free carrier mobility in the channel of 0.18 cm^2/Vs. The total drain current remains unfortunately very low due to the very low free carrier mobility of the involved organic materials. But this architecture is also transferable to polycrystalline silicon thin film transistor (poly-Si TFT) technologies for which the process can be easier and the free carrier mobility much higher than organic material one.

The first vertical polycrystalline silicon TFTs were fabricated and designed with promising results (14). For this technology, the expected strong increase of the equivalent current density per substrate area appears as a main interesting result. However, some improvements are needed, mainly the decrease of the off current for reliable applications. We will see that the approach is very close to the ULSI improvement one.

Strong increase of the width to length ratio in TFTs

Previous papers highlighted the influence of the process and the quality and the nature of the channel material on the electrical properties of the TFTs, mainly equivalent mobility of carriers in the channel, I_{ON}/I_{OFF} drain current ratio, and subthreshold slope (1). The previous studies involved planar process that means channel parallel to the substrate surface. Many approaches were made to increase the field effect mobility of the free carriers in the channel, more especially by developing monolayer structure and laser crystallization (17-18). However, the total area occupied by the TFT remains relatively high. Figure 4 shows a schematic cross-section of TFTs fabricated in this technology. An evaluation of the minimum size of the TFT with an alignment parameter, λ, leads to a minimum total area occupied on the glass substrate of 30 λ^2, with a ratio (W/L), channel width on channel length, of 1 (*see* 19). Figure 4 gives evidence that for a four mask process, the minimum size of the final TFT including source, gate, and drain contacts reaches more or less 15 λ in length and 2 λ in width. By using the same design rule, figure 5 shows a cross-section of a potential vertical TFT. The drain, channel region, and source layers are stacked and grown in the same run by adjusting the doping level by controlling the injection of the doping gas flow in the reactor (14).

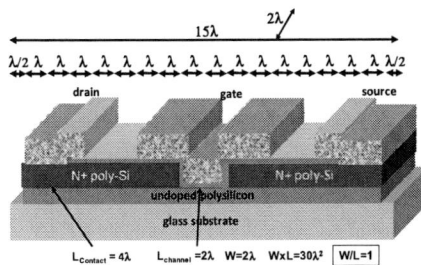

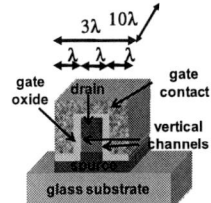

Figure 5. Cross section and size of a planar thin film transistor. Using a design rule parameter, λ, the minimum area of the transistor reaches in this case $30\lambda^2$.

Figure 4. Cross section and size of a planar thin film transistor. Using a design rule parameter, λ, the minimum area of the transistor reaches in this case $30\lambda^2$ (19).

In this case, the size of the transistor is much lower than the planar device one. With the assumption of vertical walls on both sides of the channel region, and defining the width of this region to obtain the same substrate area, the total equivalent width of the channels is in this case of 16 λ. Assuming the thickness of the non-intentionally doped layer in the range of the micrometer, with the same design kit than the planar structure, the equivalent length is approximately three times smaller. The final result gives a width-on-length ratio (W/L) about 48. It is clear that theoretically, the current driven through the structure can be about 50 times higher.

Fabrication of vertical TFTs

The fabrication process was already described on its principle in previous works (14). These studies revealed that the critical process step is the etching of the drain and channel layers in order to define the width of the source and channel regions. The resulting sidewalls become the lateral gate surfaces.

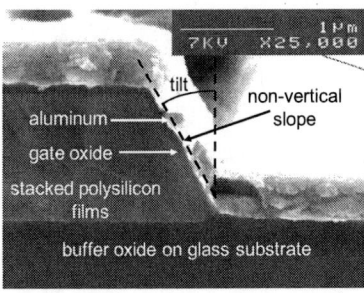

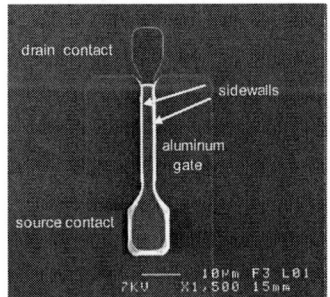

Figure 6. SEM cross-section of quasi-vertical TFT. The channels are not vertical but tilted to insure a good coverage of gate oxide and gate aluminum layers.

Figure 7. Top view of a quasi-vertical TFT (one finger or one tooth). The channel width can be high for several fingers in parallel (14).

After many experiments of dry etching followed by the deposition of the gate oxide, the best results were obtained with a tilt of the sidewalls. This "non-fully vertical" solution allows a better coverage of the gate oxide and a better continuity of the aluminum gate contact as demonstrated in (14). Figure 6 shows the best compromise. This point justifies the name quasi-vertical. The width of the finger is a little bit symmetrically enlarged but the difference is not really significant. Figure 7 shows a top view of the fabricated transistor. It is in fact a long finger or tooth. The design can gather several teeth in parallel in order to increase the drain current.

Electrical characteristics and modeling of quasi-vertical TFTs

Figure 8 shows the transfer characteristics of a quasi-vertical TFT. The shape of this curve indicates a good transistor effect, a field effect mobility of about 10 cm^2/Vs. This value is lower than the planar TFT counterpart fabricated in the same conditions. This indicates some defects in the channel region or on the quality of the interface between deposited oxide on the sidewalls and non-intentionally polycrystalline silicon channel layer. On the basis of the first electrical characteristics acting as data library, the modeling of the structure was improved.

Figure 9 shows the cross-section of the elementary quasi-vertical TFT generated by Athena module of Virtual Wafer Fab tool (SSupremIV). The associated Atlas module allows the related electrical modeling.

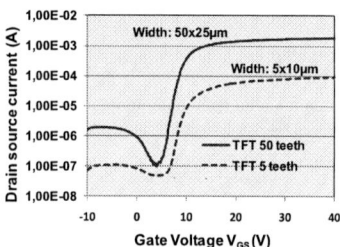

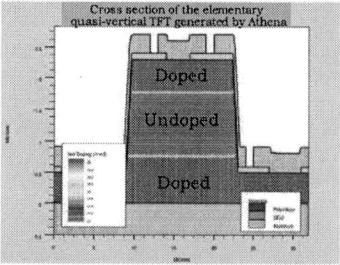

Figure 8. Transfer characteristics of vertical TFT combs with 5 and 50 teeth. The maximum direct current is higher to the milliamp. However, the I_{on}/I_{off} ratio is close to 10^4 due to a large reverse current.

Figure 9. Modeling of the quasi-vertical TFT using Virtual Wafer Fab tool (SSupremIV). On the basis of the planar TFT physical parameters, a first modeling was performed with this new architecture.

If the first results were very promising, the most important drawback comes from the reverse current that is too high and that affects the I_{on}/I_{off} ratio. It is not higher than 10^4 in these first results. This means a leakage current that flows between drain and source regions. This behavior was also predicted in the case of ULSI vertical devices; this leakage current effect was also observed in the first modeling of the integrated fully depleted SOI vertical devices (16).

34

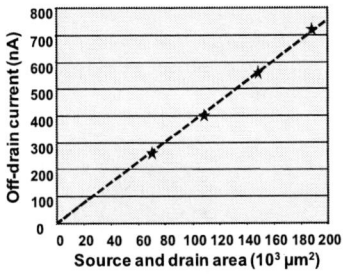

Figure 10. Variation of the off drain source current in function of the face-to-face source drain area. The off current is mainly the related leakage current between drain and source.

For the TFT, the presence of grain boundaries and defects in the channel region, non-intentionally doped, increases the leakage current. Thus, the leakage current flowing through a relatively high area is much more important in this geometry than in the case of planar devices. A special analysis of the reverse current (14) showed its direct dependence to the source and drain common area; it is proportional. Figure 10 gives evidence of this effect. Thus, the future designs of such a device must be adapted to minimize this area.

Improvement of the electrical behavior

The main problem consists to decrease the face-to-face or overlapping of source and drain. Several options are suggested i) by decreasing the area of the channel region, thanks to a decreasing of the design rules, ii) by decreasing the face-to-face area thanks to an etching of the source or drain layers, iii) by introducing an electrical barrier between source and drain layers.

The first simple idea can be the decreasing of the geometry. However this solution imposes to improve strongly the alignment limits in order to decrease the distance between the both channels but also to decrease the distance between source and drain contacts. The main advantage of this approach is an associated decreasing of parasitic capacitances.

The second way consists to block the leakage current by an electronic barrier. The insertion of an insulator between the source and drain layer can be a solution. The first idea can be to insert an oxide after the growth of the channel film, just before the deposition of the drain layer that avoids a large parasitic face-to-face zone that can be larger than the useful area. The remaining face-to-face corresponds to the gate width x channel distance area. However this proposal should lead to a significant decrease of the leakage current, but no more than a ratio three after analysis of the electrical results on fabricated test structures.

The last way was also proposed for ULSI technologies. It consists to introduce an insulating barrier between source and drain by the way of a deposited oxide.

An example of this approach was proposed to decrease the leakage current of an ULSI technology with a 50nm long channel. Figure 11 shows the presence of an insulator pocket (10). This oxide minimizes the reverse current between drain and source in reverse bias.

We have proposed a similar approach for thin film transistor. On the basis of the device shown figure 9, the non-intentionally doped region of the channels can be replaced by a deposited oxide. The new channel regions are deposited on the sidewalls with a low thickness (Figure 12). This technique allows decreasing strongly the drain source face-to-face area but also to create a similar effect as in fully depleted SOI technology. The first electrical simulations involving ATLAS and ATHENA tools confirmed this improvement.

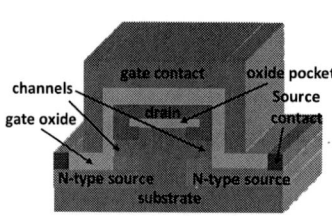

Figure 11. Expected improvement of a vertical MOSFET by including an oxide pocket that blocks the leakage current (10).

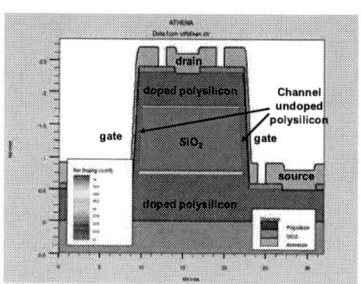

Figure 12. Improvement of quasi-vertical TFT by including an oxide between source and drain films. The channel film is just deposited on the sidewalls.

Conclusion and discussion

Many efforts are made to continue the fabulous evolution of the integration. New geometries are involved, especially 3D-technologies. The involvement of the third dimension allows increasing the width of channels without increasing of the substrate area. Many solutions are proposed and among them channel vertical sidewalls. For these vertical sidewalls, the most promising geometry is a vertical conduction, with both vertical channels facing each other. In the fields of large area technologies, we propose a new quasi-vertical thin film transistor. The first runs confirmed the possibility of fabricating such a novel structure and the first results were promising. However, the same drawbacks as in ULSI technologies occur in this new geometry, a relative high reverse current in reverse regime. To minimize this effect, similar solutions are proposed, and more especially the introduction of an electrical barrier allowing a strong decrease of the leakage current. By involving a very thin active layer for the channel regions, the improvements observed on fully depleted SOI can also occur in this structure.

To summarize, it is clear that the both technologies, ULSI and TFT ones, are now cross-dependent in several aspects. The example of the leakage current effect is common to the both vertical devices. Some improvements are also expected by changing the nature of the material of the channel region. Some solutions can be proposed by involving SiGe in order to increase the free carrier mobility.

Acknowledgments

The authors want to thank all the members of the Microelectronics Group of *Institut d'Electronique et Télécommunications de Rennes*, for the main technological results presented in this paper. A special thank to Himi-Deen Touré, Nathalie Coulon and Thierry Gaillard for their significant contribution on the results obtained on the vertical

TFT process. Samuel Crand is also thanks for his work on the physical and electrical modeling of the planar and vertical TFTs.

References

1. O. Bonnaud, T. Mohammed-Brahim, *Electrochem. Soc Transactions,* **8 (1)** 51-56 (2007)
2. X. Huang, W.Chin Lee, C. Kuo, D. Hisamoto, L. Chang, J. Kedzierski, E. Anderson, H. Takeuchi, Y-K. Choi, K. Asano, V. Subramanian, T-J. King, J. Bokor, and C. Hu, *IEEE trans. on ED*, **48**, n° 5, 880-884 (2001).
3. Mathew, *et al.* (22 co-authors) *IEDM Technical Digest. IEEE International*, **5 (5)**, 713 (2005)
4. W. Zhang, J.G. Fossum, L. Mathew, *IEEE Trans on ED*, **53**, n°9, 2335-2343, (2006)
5. Shengdong Zhang, Ruqi Han, Johnny K. O. Sin, and Mansun Chan, *IEEE Trans. on ED*, 49, n° 8, 1490-1492 (2002)
6. J.P. Colinge, M.H. Gao, A. Romano-Rodriguez, H. Maes, and C. Claeys, *Technical Digest of IEDM*, P. 595, IEDM 90-595 (1990)
7. J. Moers, *Applied Physics A*, **87**, 531-537 (2007).
8. V.D. Kunz, Takashi Uchino, C. H. de Groot, P. Ashburn, D.C. Donaghy, S. Hall, Yun Wang, and P.L. Hemment, *IEEE Trans. on ED.*, Vol. 50, No. 6, (2003)
9. S.K. Jayanarayanan, S. Dey, J.P. Donnelly and S.K. Banerjee, *Solid-State Electronics*, Volume 50, Issue 5, Pages 897-90 (2006)
10. D. Donaghy, S. Hall, V. D. Kunz, C. de Groot, P.Ashburn, *Proc. of ESSDERC 2002*, Florence (Italy). pp. 499-502 (2002)
11. Yang, B. Buddharaju, K.D. Teo, S.H.G. Singh, N. Lo, G.Q. Kwong, D.L. Inst. of Microelectron., *Electron Device Letters*, **Vol.** 29, issue7, pp 791 – 794 (2008)
12. Jyi-Tsong Lin, Ying-Chieh Tsai, Yi-Chuen Eng, Shiang-Shi Kang, Yi-Ming Tseng, Hung-Jen Tseng, Po-Hsieh Lin, *Proc. of IPFA'2008,* Singapore, (2008)
13. Chuan-Yi Yang, Shiau-Shin Cheng, Tzu-Min Ou, Meng-Chyi Wu, Chun-Hung Wu, Che-Hsi Chao, Shih-Yen Lin, and Yi-Jen Chan, *IEEE Trans on ED,* **54** n°7, 1633-1636 (2007)
14. H.D. Toure, T. Gaillard, N. Coulon, O. Bonnaud, *Electrochem. Soc. Transactions*, **16 (9)**, 165 (2008).
15. G.E. Moore, *Electronics Magazine*, **38**, p.114 (1965).
16. M. Orlowski and A. Wild, *Electrochem. Soc Transactions*, 3 (6), 1, (2006)
17. L. Pichon, F. Raoult, O. Bonnaud, H. Sehil, D. Briand, *Solid State Electronics*, Vol 38, n°8 (1995) pp 1515-1521
18. O. Bonnaud, *Polycrystalline Semiconductors V-Bulk Materials, Thin Films and Devices, Solid State Phenomena,* Scitech Publ, **67-68**, 529-540 (1999).
19. O. Bonnaud, , *ECS Transactions*, **22** (1), 293-304 (2009)

38

Similarities between μc-Si TFT with very thin active layer and FD-SOI FETs

T. Mohammed-Brahim, K. Kandoussi, K. Belarbi, H. Lhermite, N. Coulon, C. Simon

GM-IETR UMR-CNRS 6164, Universite RENNES I, 35042 Rennes Cedex, FRANCE
brahim@univ-rennes1.fr

> The subthreshold slope of microcrystalline silicon Thin Film
> Transistor is shown to be highly improved when the thickness of
> the active layer is enough decreased. In the same manner, the
> threshold voltage of these TFTs with very thin active layer is
> shown to be controlled dynamically with high efficiency by a
> second gate in front of the first one. These behaviours are
> explained considering the similarities between these TFTs and
> fully depleted SOI-MOSFETs as well as with dual-gate SOI-
> MOSFETS.

Introduction

New needs for available information anywhere at any moment lead to the
development of a lot of applications such as flexible screens, intelligent textiles, radio
frequency identification (RFID) tags on plastic and bio-patches directly integrated on
organs. Such applications need the development of a new technology to fabricate
electronic circuits on any substrate that can be flexible and transparent in some case. The
main challenge in this development concerns the maximum temperature that can be
reached by such flexible and transparent substrates without too much shrinkage to make
possible a process with some needed alignments. Indeed, for the moment, the maximum
temperature reached by such substrates is 180°C. It is reached by the so-called heat
stabilized PEN (Poly Ethylene Naphtalate) developed by DuPont Teijin Films with
thermal shrinkage lower than 0.1% at 180°C.

At such low temperature, deposited silicon can be used as active material to produce
efficient electronic devices. Silicon is usually deposited amorphous or crystallized around
250°C. Amorphous silicon based transistors are presently used in large area electronics in
spite of their low electron mobility, non stability and N-type only. Efficient electronics
for signal treatment need however the production of reliable both types of transistors
leading to CMOS electronics. Indeed, CMOS electronics availability makes possible
powerful circuit-design possibilities and then many applications in the present silicon
analog as well as digital electronics. Such CMOS electronics can be fabricated only with
crystallized silicon. Well crystallized silicon has been deposited with success around
150°C by PECVD, using highly diluted silane in hydrogen with addition of argon
sometimes. Starting from this as-deposited microcrystalline silicon material, a lot of
results on low temperature fabricated N-type thin film transistors can be found. Very high
electron mobility values were reported [1-4], without stability however. Stable TFTs were
obtained by using a 50 nm thick undoped microcrystalline silicon film as the active layer
of the transistor [5].

Reaching such very low thickness means that some effects of the reduced bulk on the TFT's parameters can occur as what happened in SOI-MOSFET when very thin bulk is used. Here, theses effects are checked on microcrystalline silicon TFTs processed at T<180°C when the thickness of the active layer is reduced.

Experimental

Top-gate and bottom-gate N-type TFTs have been processed using microcrystalline silicon (μc-Si:H) films deposited in a conventional capacitively RF-PECVD chamber on Corning 1737 glass substrate. During the growth process, the deposition temperature was kept constant by monitoring the substrate temperature at 165°C. The other fixed parameters were the total gas pressure (0.9 mbar), the silane dilution, the RF power (15 W) and the inter-electrode distance (4.5 cm). 100 nm thick undoped film and 70 nm thick N-type doped μc-Si film are deposited successively. A gas mixture of 1.5 sccm silane – 50 sccm hydrogen – 100 sccm argon is used during the deposition of the undoped film. The doped μc-Si:H film is deposited using a mixture of 1 sccm SiH_4, 100 sccm H_2 and 7.10^{-3} sccm AsH_3.

The previous stack of undoped and doped films was processed to fabricate top-gate and bottom-gate N-type TFTs in the configurations presented in Figure 1 .

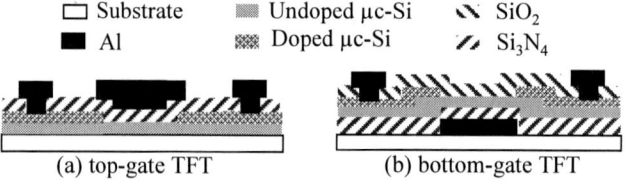

| ☐ Substrate | ▨ Undoped μc-Si | ⦚ SiO_2 |
| ■ Al | ▦ Doped μc-Si | ⫽ Si_3N_4 |

| (a) top-gate TFT | (b) bottom-gate TFT |

Figure 1. Structures of top-gate (a) and bottom-gate (b) TFTs

For the top-gate TFT, the doped μc-Si:H layer is SF_6 plasma etched to define channel, source and drain regions. Then both undoped and doped layers are SF_6 plasma etched into islands to insulate the TFTs. Afterwards, 300 nm thick silicon nitride film acting as gate insulator is deposited at 150°C in a RF-PECVD reactor using a mixture of SiH_4, N_2 and SiH_4. During the deposition the RF power was fixed at 100W and the total pressure at 0.9 mbar. The sSource and drain contacts are opened through the gate insulator by plasma etching. Finally, aluminium is deposited by thermal evaporation and wet etched to form the source, drain and gate electrodes.

For the bottom-gate TFT, aluminium is deposited first and etched to form the gate contact. Then 300 nm thick silicon nitride film acting as gate insulator is deposited in the previous conditions. Afterwards, the stack of the undoped and doped μc-Si films is deposited and etched to define the source and drain regions. Then, 160 nm thick silicon dioxide is deposited by RF sputtering without heating the substrate. The source and drain contacts are opened through this insulator. Finally, aluminium is deposited and etched to form the source and drain electrodes.

Both processes end by a thermal annealing at 180°C in an atmosphere of forming gas. This temperature is the maximum temperature reached during the processes. It is reached at the end of the process only. Before this final annealing, the maximum temperature is 165°C.

TFT's characteristics

Typical transfer characteristics of top-gate and bottom-gate transistors are shown in Figure 2 in the linear (drain-source voltage V_{DS}=1V) regime.

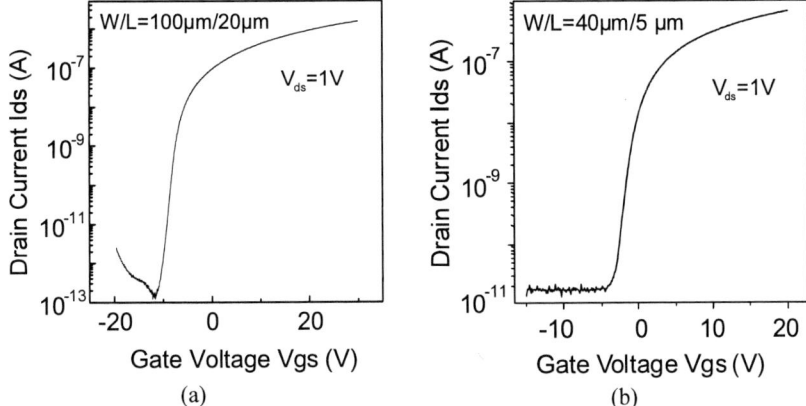

(a)   (b)

Figure 2: Transfer characteristics in the linear regime (Drain-source voltage V_{DS}=1V) of top-gate TFT (a) and bottom-gate TFT

Table I gives the main parameters of these transistors as deduced from the transfer characteristics of Figure 2. The threshold voltage is determined in the linear mode by a linear extrapolation on the gate voltage axis of the drain current Ids versus the gate voltage Vgs curve. The field effect mobility is calculated from the transconductance g_m that is the slope of the linear plot of Ids-Vgs characteristic. The subthreshold slope is the reverse of the maximum slope of the LogIds-Vgs characteristic.

TABLE I. Parameters of N-type TFT

	Threshold voltage V_{TH}	Subthreshold slope S	Mobility μ
Top-gate TFT	6.2 V	0.8 V/dec	0.5 cm²/V.s
Bottom-gate TFT	9,7 V	1 V/dec	0.33 cm²/V.s

As usual, the parameters of the top-gate TFT are slightly better than their analogous for bottom-gate TFT. Indeed, it is known that the crystalline quality of microcrystalline silicon films increases with the thickness and then saturates. The top-part of the films, where the channel of the top-gate TFTs forms, is better crystallized than the bottom-part, where the channel of bottom-gate TFTs forms.

An important characteristic of the transistors that are based on disordered semiconductors is the stability of their performance during long time functioning. The stability is improved when the structure of the material is more ordered. In this way, microcrystalline silicon TFTs are more stable than amorphous silicon transistors. Present TFTs have to be stable consequently. However, the stability depends also on the gate insulator and it is known that the stability degrades when low temperature deposited insulators are used. It is then important to check the stability of the present top-gate and bottom-gate TFTs particularly remembering that both silicon nitride gate insulator and silicon dioxide passivating insulator are deposited at low temperature. So both transistors are submitted to the same +15V gate bias stress with short-circuited drain and source during 4 hours at ambient temperature. The threshold voltage shifts by 0.16V and 0.18V for top-gate and bottom-gate TFTs respectively after this 4 hours stress. Such shift can be considered as very low highlighting the excellent stability of the present transistors.

These previous results demonstrate the reliability of the present process made at low temperature (<180°C). However, the swing around 1V/dec stays high. One way to decrease the swing is to decrease the thickness of the active layer reaching its full depletion. Indeed, the subthreshold slope was previously decreased in SOI technology when fabricating fully depleted SOI-MOSFETs. However, the electrically active defects in microcrystalline silicon due to the disordered structure can delay the depletion in the present TFTs so that the phenomenon cannot so obvious as in SOI-MOSFETs.

Decreasing the thickness of the active layer of the TFTs

To check the possibility to deplete fully microcrystalline silicon film, a simulation of the functioning of 2 transistors was done first by using SILVACO tools. Both transistors are exactly similar apart the thickness of the microcrystalline silicon active layer that is fixed at 50 nm and 500 nm respectively. The microcrystalline silicon material was considered as a homogeneous semiconductor with uniformly distributed defects. The very small size (~30 nm) of the grains in the present microcrystalline silicon film favors such consideration. The energy levels of the defects are distributed inside the forbidden energy gap of the material. The distribution is usual for disordered semiconductors with exponential band tails taking 37 meV and 39 meV for the characteristic energies of conduction and valence band tails respectively. Gaussian distribution is taken for the deep levels with 5 x 17 $eV^{-1}cm^{-3}$ as the density at the maximum. These values are determined from the transfer characteristics of N-type and P-type TFTs, simultaneously fabricated and using the same active layer. For simplicity, the same distribution of defects is taken for both transistors. This means that only the effect of the thickness of the active layer will be considered in the simulation.

Figure 3 shows the simulated transfer characteristics of both transistors. The swing is much lower for the TFT with thin active layer. Then, even if microcrystalline silicon material contains a lot of defects, the effect of the use of very thin film leads to the same result as in the SOI-MOSFETs. 50 nm thick film of microcrystalline silicon is well depleted, inducing rapid formation of the channel.

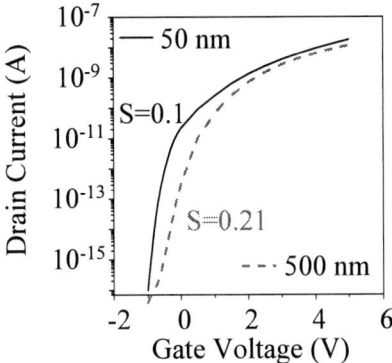

Figure 3: Simulated transfer characteristics of TFTs with 50 nm thick active layer and with 500 nm thick active layer. The subthreshold slope is only 0.1 for the first one hen it is 0.21 for the second.

To highlight more the result of the simulation, the concentration of the free electrons inside the active layers of both transistors is shown in the Figure 4 at the same positive gate voltage. The channel of thin film based TFT is pinched off when it is still horizontal in thick film based TFT.

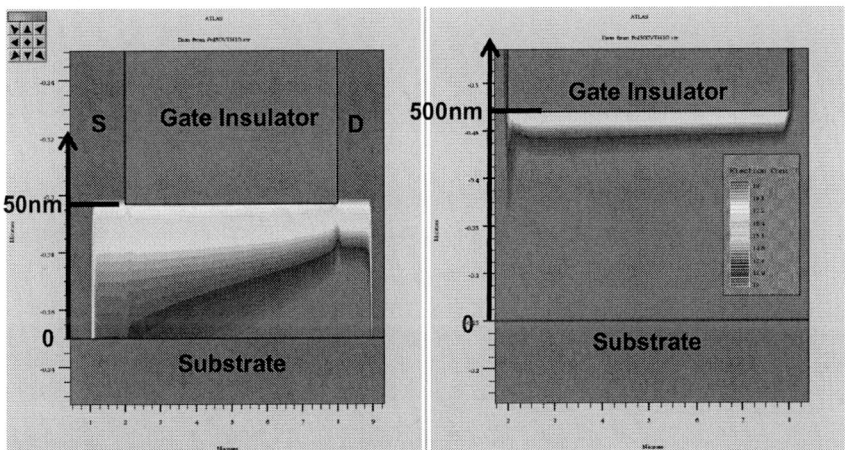

Figure 4: Concentration of the free electrons in the 50 nm thick active layer of the first TFT and in the 500 nm thick active layer of the second TFT, for the same applied gate voltage. The concentration is represented by the different colors, increasing from the purple color to the red one. The channel begins to be pinched off in the TFT with thin active layer, when it is still horizontal in the TFT with thick active layer.

The simulation shows then that very thin microcrystalline silicon film can be fully depleted when it is used as active layer of TFT, inducing then the same effect as in the SOI-MOSFETs.

Experimentally this conclusion can be confirmed by the fabrication of TFTs with very thin and thick active layer. Same transistors are fabricated following the same process of the previous part. In the first process, 100 nm thick microcrystalline silicon filmr is used as active layer. In the second, 30 nm thick film is used. Figure 5 gives the transfer characteristics of both transistors plotted in the linear regime (drain-source voltage = 1V). The swing of the thin film based TFT is nearly the half of that of the thick film based TFT that confirms the important effect of the reduction of the thickness of the active layer in the improvement of the swing.

The mobility is however lower. Its value is close from that of the previously presented bottom-gate TFT. Indeed, as for the bottom-gate TFT, the channel of the thin film based top-gate TFT forms inside the 30 nm thick film that is less crystallized than thick film.

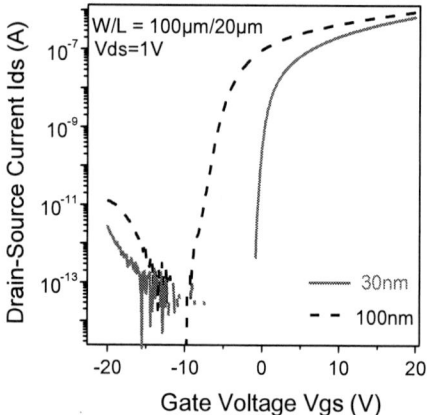

Figure 5: Transfer characteristics of TFTs with 30nm and 100 nm thick active layer in the linear regime. The transistors have 100 μm channel width and 20 μm channel length.

TABLE II. Subthreshold slope S and Field effect mobility of the N-type TFTs whose the transfer characteristics are presented in Figure 5.

Thickness of the active layer	Subthreshold slope S	Mobility μ
30 nm	0.6 V/dec	0.39 cm^2/V.s
100 nm	1.1 V/dec	0.48 cm^2/V.s

Dual-Gate TFT

The other parameter that it has to be well controlled is the threshold voltage. Indeed, the effect of any charge inside the gate insulator or in the substrate can be very important when using very thin microcrystalline silicon film as active layer. Of course, the threshold voltage can be controlled by the reproducibility of the process. However, it can be useful to control dynamically the threshold voltage during the functioning and after the fabrication. Such control can be made by putting a second gate in front of the main gate, fabricating then a dual-gate transistor.

Dual-gate transistors were first developed using SOI technology [6, 7] taking advantage of the very thin layer to enhance the interaction between both gates. The coupling between both gates is well known now in this technology. However, the functioning of TFTs is slightly different from SOI-FETs. Indeed, the active layer where the channel forms is undoped in silicon TFTs. In fact, undoped microcrystalline silicon films are slightly N-type doped. The drain-source current in the off-regime is limited only by the high resistivity of this undoped film. N-type TFTs works in the electron accumulation regime. For SOI-FETs, the active layer is P-type doped and the transistor works in the inversion regime. This difference between TFTs and SOI-FETs can lead to some difference in the appearance of the coupling effect.

As described in the experimental part a usual bottom-gate TFT is fabricated. It uses 300 nm thick silicon nitride as insulator of the back main gate and 50 nm thick undoped microcrystalline silicon film as active layer. The present process differs from the previous one only in the last step where the final aluminum film is etched to define source and drain contacts as previously, but also to define the contact of the additional top-gate.

To check the efficiency of the control of the threshold voltage of the bottom-gate TFT by the additional top-gate, its transfer characteristic is plotted at different fixed applied voltage on the top-gate V_{GTOP}. Figure 6 presents these characteristics at different V_{GTOP}, varying from negative to positive values.

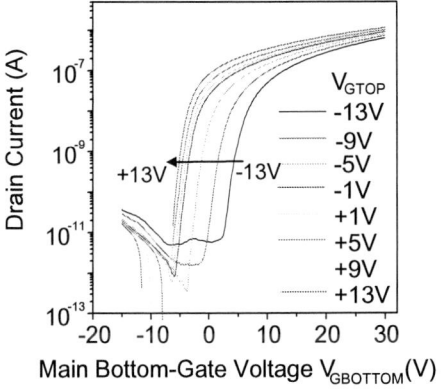
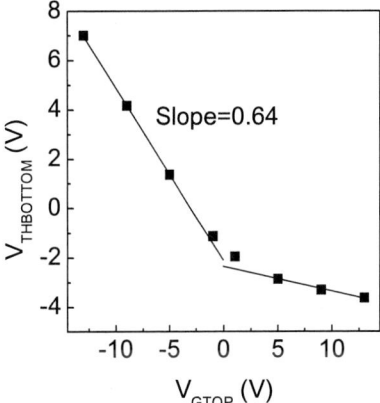

Figure 6 : Transfer characteristics in the linear regime (Drain-source voltage = 1V) of the bottom-gate TFT for different values of the voltage applied to the top-gate, varying from -13V to +13V.

Figure 7 : Threshold voltage $V_{THBOTTOM}$ of the bottom-gate TFT as a function of the voltage V_{GTOP} applied to the top-gate.

The characteristics shift well with the voltage applied to the top-gate at least when this voltage is negative. Indeed, the shift seems to saturate for the positive values of V_{GTOP}. Into quantifying the efficiency of the modulation of the threshold voltage $V_{THBOTTOM}$ of the bottom-gate TFT by the voltage applied to the top-gate V_{GTOP}, $V_{THBOTTOM}$ is plotted as a function of V_{GTOP} in Figure 7. The plot shows a linear dependence of $V_{GBOTTOM}$ with negative V_{GTOP}. The slope is 0.64. When V_{GTOP} is positive the dependence is very weak and the shift tends to saturate.

This behavior is similar to that often reported for dual-gate SOI-MOSFETs (see for example the figure 3a in [6]). The present coupling of 0.64 is in the order of what it is given in this reference, even if the thickness of the active layer is much lower in this reference. This comparison highlights the high quality of the coupling obtained here with thin microcrystalline silicon films.

The shift of $V_{THBOTTOM}$ with negative V_{GTOP} can be explained from the intrinsic functioning of N-type TFT. As the active layer is non-intentionally doped and in fact slightly N-type and considering its very low thickness, it is fully depleted when V_{GTOP} is negative. In this case, both gates are strongly coupled and V_{GTOP} influences greatly the threshold voltage of the bottom-gate TFT. When V_{GTOP} becomes positive, an accumulation of electrons occurs near the top interface. The coupling is highly reduced and destroyed for high positive value of V_{GTOP}. The threshold voltage of the bottom-gate TFT is less dependent on V_{GTOP} and then no longer depends on it for high positive values.

Such decoupling between both gates is also encountered in FD-SOI MOSFETs for large negative values of the parasitic gate voltage when the interface near this gate is fully accumulated by holes [7]

Conclusion

An engineering of thin film transistors fabricated at low temperature that is compatible with the use of plastic substrate was presented. The development of a new technology needs to insure for the reliability of the devices and particularly of the basic element that is the transistor. It needs also to check all the possibilities to improve the performance taking into account the own limitations of the technology. Here, we insured first for the stability of the TFTs with the gate located at the bottom or on the top of the active layer.

Then the improvements presented here followed the experience of what it was made before with the SOI technology. Using very thin active layer led to an improvement of the subthreshold slope. Using dual-gate TFT led to a very efficient control of the threshold voltage by a voltage applied to the second gate. This dynamic control of the threshold voltage will be very important in the pairing of transistors leading to a CMOS technology for example.

Following what it was made with the SOI technology led us to compare both technologies that are similar with important differences in the functioning however.

Finally, the main following improvement will be the increase of the field effect mobility that is low, limiting then the functioning frequency. However, the present devices work at some kHz that it is sufficient to be integrated with some sensors working in the frequency range of the human activities.

Acknowledgments

The work was made in the framework of the French project "eFlexSi" ANR-09-blan-0163.

References

1. I-C Cheng, S.Wagner, *Thin Solid Films* **427**, 56 (2003)
2. A. Saboundji, N. Coulon, A. Gorin, H. Lhermite, T. Mohammed-Brahim, M. Fonrodona, J. Bertomeu, J. Andreu, *Thin Solid Films* **487**, 227 (2005)
3. C.H. Lee, A. Sazonov, A. Nathan, *Appl. Phys. Lett.* **86**, 222106 1-3 (2005)
4. K. Kandoussi, C. Simon, N. Coulon, T. Mohammed-Brahim, *Proceedings of Active Matrix Flat Panel Displays (AM'FPD 06)*, Tokyo, 35 (2006)
5. K. Belarbi, K. Kandoussi, N. Coulon, C. Simon, R. Cherfi, A. Fedala, T. Mohammed-Brahim, *ECS Trans.*, **16**, 121 (2008)
6. F. Balestra, S. Critoloveanu, M. Benachir, J. Brini, T. Elewa, *IEEE Electron. Dev. Lett.* **8**, 410 (1987)
7. Y. Liu, M. Masahara, K. Ishii, T. Sekigawa, H. Takashima, H. Yamauchi, E. Suzuki, *IEEE Electron Dev. Lett.* **25**, 510 (2004)
8. M. Masahara, Y. Liu, K. Ishii, K. Sakamoto, K. Endo, T. Matsukawa, K. Ishii, T. Sekigawa, H. Yamauchi, H. Tanoue, S. Kanemaru, *IEEE Trans. Electron Dev.* **52**, 3046 (2005)

48

MILC PMOS POLY-SILICON TFT CIRCUITS AND APPLICATION IN SOP

Pengfei Sun[a], Shuyun Zhao[a], Tsz-Kin Ho [a], Zhiguo Meng [b], Man Wong [a], Hoi Sing Kwok [a]

[a] Center for Display Research, Department of Electronic and Computer Engineering, Hong Kong University of Science and Technology, Clear Water Bay,   Kowloon, Hong Kong, China
[b] Institute of Photo-electronics, Nankai University, Tianjin, China

Research towards thin film circuits is one of the most attractive directions of the large-area electronics and flexible electronics. In this work, the research on TFT shift register circuit which is critical for system integration on panel (SOP) is presented. The whole circuit is composed of MILC PMOS poly-Si TFTs. The TFT device exhibited field-effect mobility of $65.21cm^2/Vs$, threshold voltage of -3.5V and sub-threshold swing of 0.56V/dec. Some special design considerations were adopted to improve the robustness of the circuits. The whole shift register is composed of as many as 180 stages and exhibited well performance with frame frequency from 22Hz to 220Hz at the power supply voltage of 11V. No signal attenuation and distortion appeared from the first to the final stage. The effect of bootstrapping and dynamic behavior is analyzed. The fabrication and reliability analysis are also discussed in this paper. Overall, the high performance driver circuits based on our MILC PMOS process are possible and have the potential in the application of SOP.

Introduction

TFT circuits have been studied widely because of the flexible and large area electronic applications for the last several years (1).

There are many options which are polycrystalline silicon (poly-Si) TFT, amorphous silicon (a-Si) TFT, organic TFT and single crystalline silicon to realize TFT circuit (2). For a-Si silicon and organic TFT, there are some intrinsic limitations concerning integration of larger scale circuits, as a result of low mobility and high threshold voltage. During these years, several attempts of transferred single crystalline Si layers on glass substrate are reported (3). Also, some of recent literatures show that single-grain (SG) Si-TFT has the potential to develop a large scale digital and analogue circuit system by the special process (4).

One of the most important concerns for TFT circuit is process variation and fabrication cost. To integrate high performance circuits of TFT devices, the low temperature poly-Si (LTPS) technology is still the most widespread. Metal induced lateral crystallization (MILC) is a promising technology to realize high performance p-type poly-Si TFT. However, due to the intrinsic grain boundaries of poly-Si causing a negative effect on device performance such as mobility and uniformity, there is always much difficulty and little progress to realize high performance circuits using such simplified process. In this work, the research on large scale TFT shift register circuit which is critical for system integration on panel (SOP) is presented. The proposed design logic uses all PMOS MILC TFT. The process flow and device performance which is relative to our previous work is briefly introduced. The design considerations will be presented and discussed. The measurement results from the circuits we fabricated will be shown and analyzed.

MILC Poly-Silicon TFT device fabrication

In our previous works (5), the MIC/MILC poly-Si TFT performance has been improved and optimized continuously.

The fabrication process in this study began with 4-inch Eagle 2000™ glass substrate covered with 300nm low temperature oxide (LTO). Then 45nm a-Si was deposited by low-pressure chemical vapor deposition (LPCVD). 100nm LTO was deposited by LPCVD too after a-Si. Then, Boron ion was implanted to adjust the threshold voltage of the TFT. After the induce hole was formed by etching the former LTO, 4nm nickel/silicon oxide mixture was deposited by DC sputtering. Then the a-Si was annealed at 590°C for 2.5 hours. After removing the nickel and LTO layer, the active layer was ready for device fabrication. Figure 1 shows the MILC process and cross-section of the active layer after crystallization.

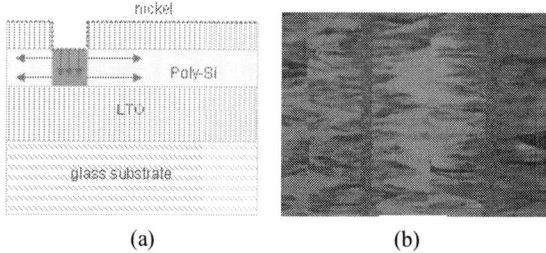

| (a) | (b) |

Figure 1. (a) MILC process   (b) The cross-section of poly-Silicon

The active channel was formed by photolithography and Freckle wet etching. After photo-resist removal, 100nm LTO was deposited by LPCVD as gate insulator. After that, 200nm metal was deposited and etched to form gate. Boron was implanted with energy of 40Kev and dosage of $4E15/cm^2$. After 500nm LTO was deposited, contact hole was opened and metal electrode was sputtered. The device process was finished after FGA about half an hour. The device fabrication and characteristics are shown in Figure 2.

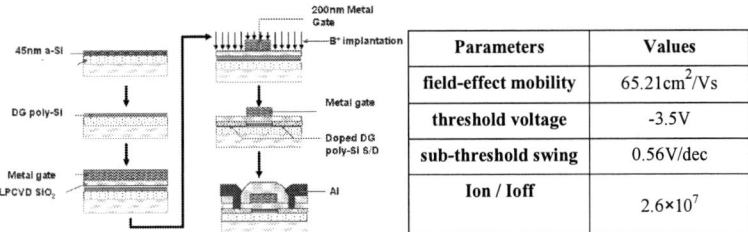

Parameters	Values
field-effect mobility	$65.21 \text{cm}^2/\text{Vs}$
threshold voltage	-3.5V
sub-threshold swing	0.56V/dec
Ion / Ioff	2.6×10^7

Figure 2. Device fabrication and characteristics

The TFT logic circuit design consideration

There are various circuit structures to implement the function of shift register (6). In order to fit in our MILC process, the circuit structure is proposed and modified based on some conventional circuit widely used in industry. Figure 3 shows the schematic of one stage of shift register and the timing of the signal waveform. As shown in schematic, P2 acts as switch transistor and P5 acts as driver transistor which is similar to the conventional TFT-AMOLED pixel circuit.

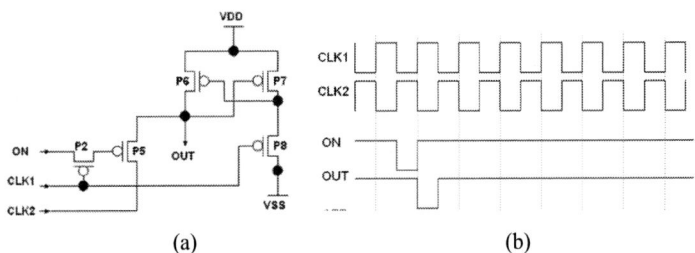

(a)   (b)

Figure 3. (a) The schematic of single-stage circuit (b) The timing signals

Transistor P2 will be turned on periodically, when the ON signal stays at high level, the gate of transistor P5 is charged positively and repeatedly. While the ON signal stays at low level, the gate of transistor P5 is charged negatively. In this case, transistor P5 stays at the open state dynamically. Then the OUT signal will be generated by CLK2 through the driver transistor P5. Transistors P6, P7 and P8 execute the output voltage storage function which is similar as simplified DRAM circuit (7). The W/L ratio of each transistor could be optimized by Smart spice EDA tool.

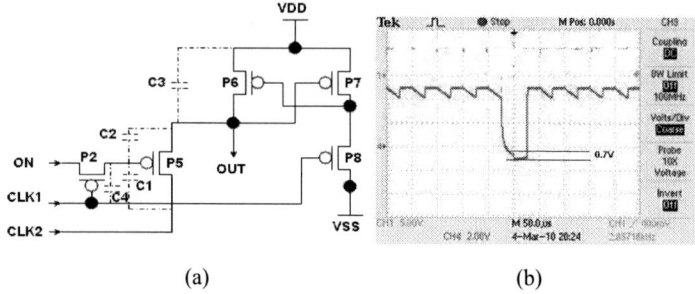

(a)   (b)

Figure 4. (a)The schematic considering parasitic capacitance
(b) The voltage drop caused by the bootstrapping effect

$$\Delta V_{GS} = \frac{C1}{C1+C2+C3+C4}\Delta V_{CLK2}$$

Figure 4 shows the bootstrapping effect occurring when the P5 driving function would be executed. Because of the coupling between CLK2 wire node and gate node of P5, the dynamic charge may be reconstructed by coupling with other nodes. Proper bootstrapping effect would be advantageous to sharp the falling edge of the OUT waveform. However, it has adverse effect on the gate oxide of P5 because too much charge trapped would cause breakdown.

In this work, the voltage drop by bootstrapping was optimized to approximate 0.7V. Figure 4 also shows the voltage drop waveform in the experiment.

The shift register layout was designed to be best compatible with MILC poly-Si TFT process. Shift register cells were carefully tested and their performance were evaluated and compared with the goal of realizing the optimum layout structures suitable for poly-Si TFT driver system development.

It has been reported commonly used TFT shift registers have robust performance but the topologies are not compatible with MILC process to achieve high stability and large scale. In our work, split gate and split channel structures were carefully considered and tested to reduce the negative effect of poly silicon's poor uniformity (8). Splitting the large transistor into small sub-transistor which has constant channel width and length can improve the reliability and accuracy of the design aim during the process. Also, the thickness of gate oxide

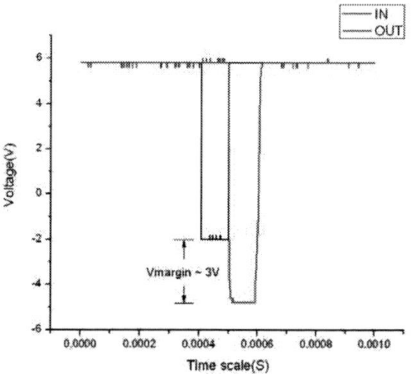

Figure 5. The noise margin analysis

which acts as dynamic storage capacitance was optimized in process to trade off the bootstrapping effect.

Due to the lower mobility and higher leakage compared with single crystal silicon, the characteristics of TFT device present some deduction and distortion in the shift register outputs. In this work, the issue of noise margin has been carefully tested. The "DRAM" structure composed of P6, P7, and P8 can hold the high level of output well. The Figure 5 shows the low level noise margin when IN was stimulated by pulse with smaller gap between high and low levels. The result shows that the noise margin can achieve around 3V. So, if the output of previous stage is distorted to less than 3V, the pulse can also be transmitted with no deduction and the whole serious shift register can also work well.

Results and discussion

The system block diagram of the shift register circuit, the microscopic photograph of 180-stage shift register fabricated, output waveforms probed by a four channel oscilloscope are shown in Figure 6. The output pulse signal is transmitted series and the circuit can implement the function as designed.

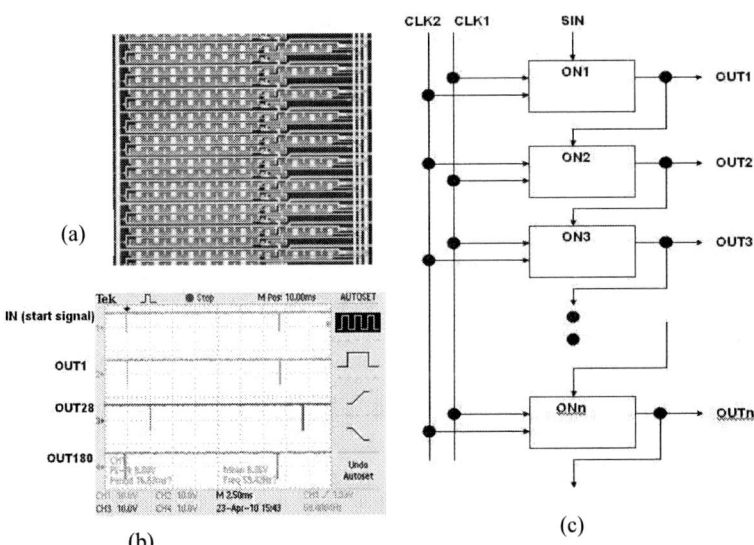

Figure 6 (a) The microscopic photograph of the circuit.
(b) The timing result probed by 4 channel oscillator.
(c) The system block diagram.

The circuit performance under different frame frequency has been tested. The two-phase-clock frequency is proportional to the frame frequency and has little effect on the falling time of each pulse. However, due to the topology of the circuit, the rising time is affected by the clock frequency because the clock signal indirectly drives the transition from low level to high level.

The waveforms specialized on the falling and rising time are shown in Figure 7. Table I summarizes the signal edge timing. In the normal frame frequency of 60Hz, the falling time and rising time are 2.5μs and 5.6μs respectively. The ratio of edge time on duty is 8.75%. The rising time under the low frequency is larger than that under the high frequency while the falling time varies little.

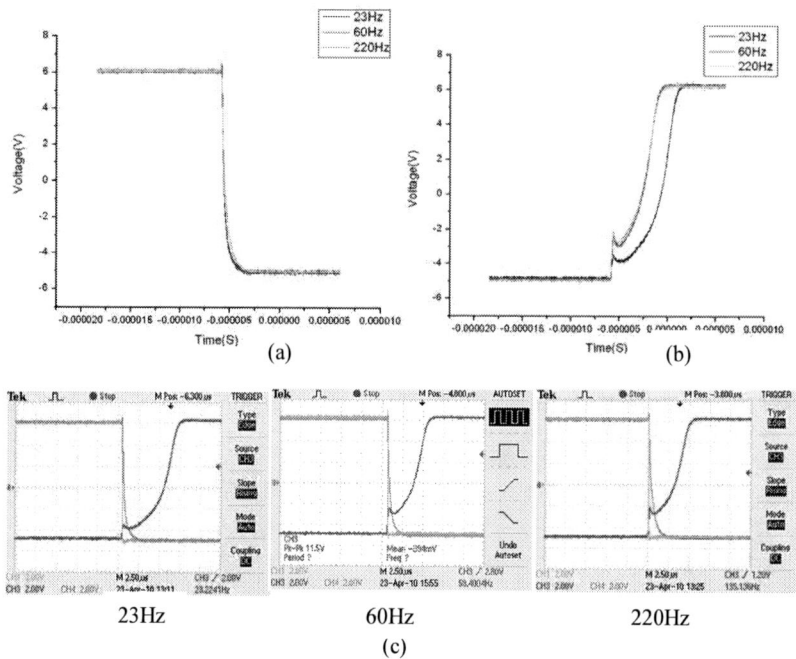

Figure 7. The falling time (a) and rising time (b) under different frame frequency; (c) The waveform captured from oscillator

Table I. The timing analysis

Frame Rate	Clock Frequency	Falling time	Rising time	The ratio of edge time on duty
23Hz	2.07KHz	2.5μs	7.5μs	4.14%
60Hz	5.4KHz	2.5μs	5.6μs	8.75%
220Hz	19.8KHz	2.6μs	5.5μs	32.08%

Summarizations

The TFT shift register circuit by MILC process has been investigated in this work. By our special design consideration and fabrication, the experiment results demonstrate the 180-stage shift register circuit has robust and high performance at speed, noise margin and device uniformity. The TFT circuits based on our MILC PMOS process can satisfy the requirement of SOP and have the competitive potential in the application of SOP display.

Acknowledge

This work was supported by the Hong Kong Government Research Grants Council under Grant Number AOE/P0308PG2.

References

1. Nathan, P. Servati, K.S. Karim, et al, *IEE Proc.-Circuits Devices Syst.,* Vol. 150, No. 4, 329-338 (2003).
2. Jae Ik Kim a, Jae Won Choia, Wonjae Choi, et al, *Solid-State Electronics,* 54, 299-302 (2010).
3. Xuejie Shi, K.Henttinen, T.Suni, et al, *IEEE Electron Device Lett,* Vol.24, No.9, 574-576 (2003).
4. R. Ishihara, A. Baiano, T. Chen, et al, *ECS Transactions,* 22, (1), 57-68 (2009).
5. Shuyun Zhao, Zhiguo Meng, Xuedong Li, et al, *SID 07 DIGEST,* P-17 (2007).
6. Shu-Hun Yang, Wei-Cheng Chen, Hsiao-Yi Lin Ching-HoneLee,et al. *SID07,* 245-248 (2007).
7. Ching-Hone Lee, Hsiao-Yi Lin, Chaung-Ming Chiu, et al. *IDW/AD '05,* 1027-1030 (2005).
8. Kris Myny, Monique J. Beenhakkers, Nick A. J. M. van Aerle, et al, *2010 IEEE International Solid-State Circuits Conference, SESSION 7,* 140-142 (2010).

56

Source-Gated Transistors for
Versatile Large Area Electronic Circuit Design and Fabrication

R. A. Sporea[a], X. Guo[b], J. M. Shannon[a] and S. R. P. Silva[a]

[a]Advanced Technology Institute, University of Surrey, Guildford, GU2 7XH, U. K.
[b]Displays and Lighting Centre, Shanghai Jiao Tong University, Shanghai, 200240, China

> Source-gated transistors (SGTs) comprise a blocking contact or
> potential barrier at the source, which control the current. The paper
> describes how SGTs can be optimized for particular applications
> and for specific semiconductor material systems. It is shown how
> the saturation voltage can be designed to be an order of magnitude
> smaller than in equivalent FETs to give power savings of over 50%
> for the same current output. The SGT also achieves a better
> saturation regime, with lower output conductance over a larger
> range of drain voltages. Flat-panel lighting, remote sensing and
> signal processing and large-area circuits made using inexpensive
> but imprecise patterning techniques are some of the applications
> which could benefit from incorporating these devices.

Introduction

Improvements in material science and fabrication techniques are rapidly increasing the speed, analog capabilities and cost-effectiveness of complex large area electronic (LAE) systems. For future power-efficient, robust applications, similar gains can be foreseen by advances in device engineering.

The source-gated transistor (SGT) (Figure 1) is a type of thin-film transistor in which the current is mainly controlled by the effective height of a source barrier [1]. The gate lies opposite and overlaps the source and represents the means through which the source barrier height can be controlled during operation. SGTs can be produced in any technology which permits staggered electrode configurations and is compatible with semiconductor systems for which a potential barrier can be created at the source. There are several routes to engineering this barrier, the most convenient of which is the Schottky (rectifying metal-semiconductor) contact [2, 3].

Due to their structure, the operation of SGTs [1, 4] is quite different from that of conventional thin-film field effect transistors (FETs). However, SGTs are compatible with FET technologies and both types of devices can be used in a given design, exploiting the advantages of each, as required.

In this paper we examine the key areas in which SGTs have the potential of superior performance in comparison with conventional FET devices of similar geometry and operating under comparable conditions. The discussion begins with an enumeration of essential performance aspects and an explanation of how the design of the SGT allows improvements in these specific areas, and we conclude with a description of the manner in which these advantageous features can be applied to different material systems and common circuit functions.

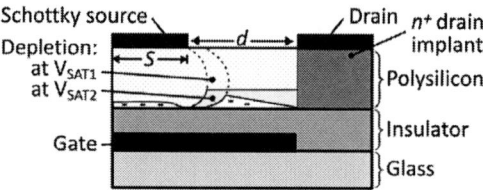

Figure 1. Schematic cross-section of a SGT in which the drain is self-registered to the gate, showing: the parasitic FET channel which forms between source and drain, source length (S), source-drain separation (d) and depletion region when the FET channel pinches off at the source ($V_{DS}=V_{SAT1}$) and at both the source and the drain ($V_{DS}=V_{SAT2}$).

Main Performance Characteristics of the Source-Gated Transistor

The presence of a reverse biased potential barrier at the source [1 - 4] has several effects on the operation of the device. Specifically, because the current through the entire transistor is limited by the source barrier, it will be lower than that developed by a FET with the same geometry and bulk doping. The current will also have a positive temperature coefficient [5] and will not be influenced significantly by some of the parameters which define the operation of a FET, such as the size of the source-drain gap (d, Figure 1). There are a number of advantages of the SGT architecture as well; a brief description of the main differences between SGT and FET operation follows.

Saturation voltage

In FETs, the current saturates at a drain voltage equivalent to the difference between the gate bias and the threshold voltage. As such, a 1V increase in gate voltage will raise the saturation voltage by 1V. In comparison, the saturation voltage of SGTs is governed by the electrostatics of the stack formed by the source electrode, semiconductor, insulator and gate electrode [6]; if a thick insulator or very thin semiconductor are used, then the change in saturation voltage with SGT gate voltage can be significantly lower than that of a FET (Figure 2). For a given drain current the power dissipation in any transistor increases proportionally with the serial voltage drop on the device and a lower saturation voltage translates directly in power savings. In Figure 2, the SGT draws around 60% less power than the FET at the onset of saturation.

The drawback of low saturation voltage is the reduced value of transconductance, and it occurs as a consequence of increased insulator thickness, which effectively lowers the gate field for a given gate bias.

Output impedance in saturation

As the barrier controls the current in saturation, the operating regime of the parasitic FET which forms between the source and the drain becomes less important. Strong saturation can be achieved well below $V_D = V_G - V_T$. When the parasitic FET is in saturation, the characteristic becomes extremely flat, which is advantageous for high-impedance active loads, current sources and analog amplifiers. Furthermore, in high mobility semiconductor technologies, the source barrier greatly reduces the kink effect [4] by extracting minority carriers and preventing bipolar amplification [7].

Overall, this means that the flat saturation region is increased toward both lower and higher voltage compared to that of a regular FET for improved dynamic range, and it is worth noting that this may lead to additional improvements in energy efficiency by not allowing the current to increase significantly with drain voltage.

Operating frequency

Another consequence of the control of current by the source barrier is that the distance between the source and the drain plays no role so long as it is not large enough to make the FET channel too resistive. The source-drain gap can be made very small, giving a higher operating speed to the device while maintaining a flat saturated current. The operation of the gate-source stack also leads to current crowding at the edge of the source, which means that the source length can be decreased for improved source current density. It has been shown that the operating frequency is proportional to average current density [8]; low-mobility semiconductor technologies could greatly benefit [9], especially in the realm of analog design, as the high output impedance characteristic is retained.

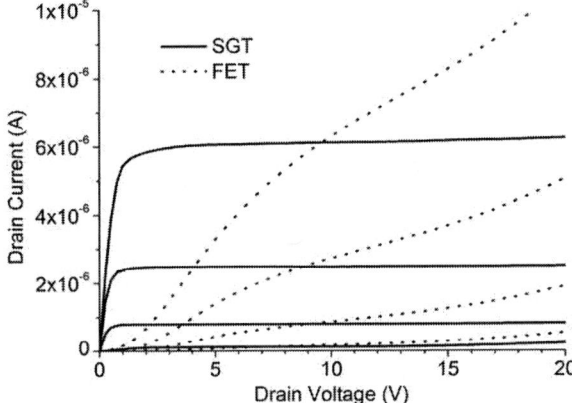

Figure 2. Output characteristics for SGT and FET devices in polysilicon with identical geometries and biased for similar saturated currents. The lower saturation voltage and flatter saturation characteristic of the SGT are apparent. Owing to lower transconductance, to achieve the same drain current, the drive voltage of the SGT is higher than that of the FET. Top to bottom gate voltages: –3, –4, –5, –6 V (FET); 20, 9, – 1, –3V (SGT).

Tolerance to process variations and suitability to cost-effective patterning techniques

The SGT architecture is also well suited for emerging technologies such as inkjet printing and stamping, where fabrication tolerances may be high. As seen in Figure 3, the current is independent of source-drain gap and, due to current crowding at the source edge, can be made to be very insensitive to source length [10]. This is a useful feature which can improve consistency of performance across a large-area substrate and, to some extent, the reliability of complex analog circuits realized with these technologies.

Off-current

The minimum off-current through a FET is strongly dependent on bulk doping; SGTs can have a much lower off-current: the reverse leakage through the source barrier [2, 3, 5]. A diagrammatic representation of this effect is given in Figure 4.

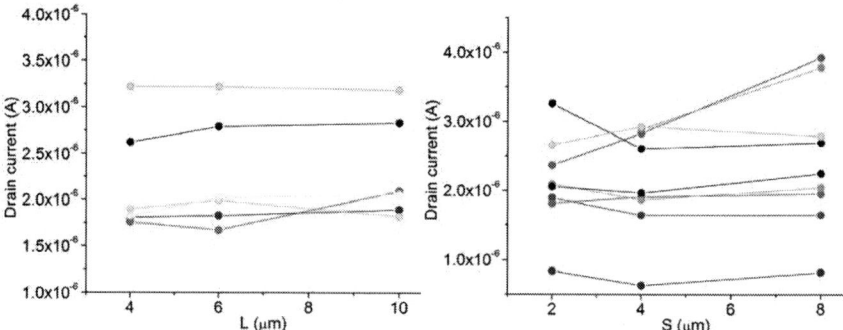

Figure 3. Drain current measured on poly-Si SGTs of various sizes biased at roughly the same current, showing little change with: *a)* source-drain gap and *b)* source length.

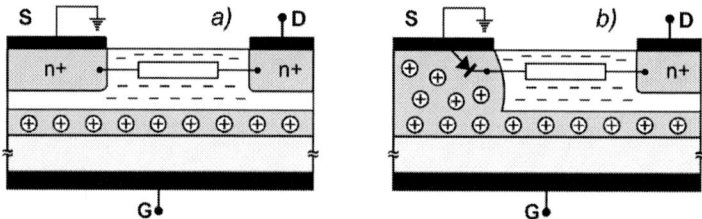

Figure 4. Schematic diagram explaining why an n-type FET (*a*) realized in a strongly n-type semiconductor does not turn off well but a similar SGT (*b*) does.

Stability under electrical stress

During SGT operation, the number of excess carriers in the device is dictated by the operating conditions of the source barrier and, as such, can be substantially smaller than in a similar FET. In amorphous semiconductors especially, this produces a large improvement in electrical stress-induced threshold voltage instability [10], as it has been proven that defect generation is related to excess carrier density [11].

Temperature coefficient of drain current

When high Schottky barriers are used, the current through the device is strongly temperature dependent [2, 3, 5]. On the one hand, this effect can be turned into a useful feature by applying the high barrier SGT as a very sensitive temperature sensor. On the other hand, the temperature dependence can be minimized by designing lower source barriers. Low barrier polysilicon SGTs have shown positive temperature coefficients in the order of those of FETs [5]. A trade-off between the quality of saturation and the temperature coefficient must be made in choosing the specifications of the source barrier.

Benefits of Source Gated Transistors to
Common Applications and Material Systems

In the previous section we have shown how the operation of the SGT, and specifically the existence of a reverse-biased potential barrier at the source, leads to some improvements in key parameters over conventional FET structures. Several common circuit applications (Table I) could benefit from incorporating SGTs.

For current sources and active matrix pixel drivers, the low saturation voltage translates into low supply voltage operation for greater energy efficiency, and the flat saturation characteristic over a wide range of voltage improves signal amplification and reduces crosstalk through the power lines.

Low barriers can be used to ensure low temperature coefficients; conversely, the SGT can function as a high-resolution sensor.

The improved stability to electrical stress is advantageous to all analog applications, reducing the need for compensating schemes and lowering circuit complexity.

As switches, SGTs may have lower off-current than similar FETs, which would improve static power dissipation. Their use could contribute to a smaller circuit footprint in mixed signal circuits such as pixel drivers: as a result of lower leakage and improved charge retention, storage capacitors can be reduced in size.

TABLE I. Common applications and their key performance characteristics which could be improved by implementing with SGTs

Performance characteristic	Current source and pixel driver	Analog gain stage	Temperature sensor	Digital switch
Output impedance in saturation	●	●	●	
Operating frequency		○		○
Saturation voltage and static power dissipation	●	●	●	
Temperature coefficient of drain current	○	○	●	
Off-current				●
Sub-threshold slope				○
Stability under electrical stress	●	●	●	

● – feature which could improve if using SGT implementation
○ – feature which does not change if using SGT implementation

Some detrimental material peculiarities can be worked around by using source-gated transistors in place of regular FETs (Table II): stability is improved in amorphous materials; the kink effect is minimized in high-mobility semiconductors and the operating frequency of organic SGT can be increased by taking advantage of the field-dependent mobility [9]. In virtually all materials the output impedance in saturation is increased and the saturation voltage can be greatly diminished leading to greater energy efficiency.

As the current through the SGT is nearly insensitive to changes in source length or source-drain gap, registration and size errors are not critical. Printing, stamping and other high-volume, cost-effective fabrication processes can be employed with minimal impact on circuit performance.

These characteristics of SGTs may open the path to analog and mixed signal circuit applications which were previously impractical to consider or too expensive to fabricate, e.g. flexible, printed sensors, displays and other analog and mixed signal designs.

TABLE II. Material systems and their key performance characteristics which would improve significantly by implementing with SGTs

Performance characteristic	a-Si:H and other low-mobility inorganic	Poly-Si and high mobility inorganic	Solution-processed organic
Output impedance in saturation	●	●	●
Operating frequency	●	○	●
Down-scaling and process variability	●	●	●
Stability under electrical stress	●	○	●

● – feature which could improve significantly if using SGT implementation
○ – feature which could improve somewhat if using SGT implementation

Conclusions

We have outlined a number of performance characteristics which make the source-gated transistor (SGT) preferable for certain circuit application to standard thin-film field-effect transistors (FETs). The versatility of the SGT would allow designs containing it to potentially improve operational parameters or circumvent undesirable properties of particular material systems: power dissipation, analog amplification, leakage current, stability and uniformity of performance over a large substrate area. Flat-panel lighting, remote sensing and signal processing are some of the areas which could benefit from incorporating these devices. SGTs and FETs can be included in the same design and the advantages of each exploited simultaneously, which may prove a great benefit when optimizing future designs for specific low-power, large area applications

Acknowledgments

The authors thank Dr. M. J. Trainor and Dr. N. D. Young at Philips Research for providing the polysilicon SGT devices. X. Guo is funded by the National Natural Science Foundation of China (No. 60906039 and No. 60910295). R. A. Sporea is supported by EPSRC through grant number EP/P503892/1. This work was undertaken as part of the International Joint Project – 2009/R3 Royal Society / China NSFC.

References

1. J. M. Shannon and E. G. Gerstner, *IEEE Electron Dev. Lett.* **24**, 405 - 407 (2003).
2. S. M. Sze, "Physics of semiconductor devices", Second Edition, 245 - 311, John Wiley & Sons, Inc. (1981).
3. E. H. Rhoderick, *Solid-State and Electron Devices, IEE Proc. I,* **129**, 1-14 (1982).
4. R. A. Sporea et al., *IEEE Trans. Electron. Devices,* **57**, 10, 2434 - 2439 (2010).

5. R. A. Sporea et al., *Proc. ESSDERC2010*, 13 - 17 (2010).
6. J. M. Shannon and E. G. Gerstner, *Solid State Electron.*, **48**, 7, 1155-1161 (2004).
7. A. Valletta, P. Gaucci, L. Mariucci, G. Fortunato and S. D. Brotherton, *Appl Phys Lett* **85**, 3113 - 3115 (2004).
8. J.M. Shannon and F. Balon, *IEEE Trans. Electron Dev.* **56**, 2354 - 2356 (2009).
9. X. Guo and J. M. Shannon, *IEEE Electron Dev. Lett.* **30**, 365 - 367 (2009).
10. R. A. Sporea, X. Guo, J. M. Shannon, S. R. P. Silva, *Trans. CAS2009*, 413 – 416 (2009).
11. J. M. Shannon, *Appl. Phys. Lett.*, **85**, 2, 326 – 328 (2004).
12. R. B. Wehrspohn, M. J. Powell, and S. C. Deane, *J. Appl. Phys.*, **93**, 5780 (2003).

64

Single-Grain Germanium TFTs

Ryoichi Ishihara, Tao Chen, A. Baiano, M.R. Tajari Mofrad and C.I.M Beenakker

Delft University of Technology, Delft Institute of Microsystems and Nanoelectronics (DIMES), Delft, The Netherlands

We review our recent achievements in location-control of Ge grains and high performance single-grain (SG) Ge thin film transistor (TFT) fabricated inside a Ge grain. Large Ge grains having a grain size of 10 μm were obtained at predetermined positions by the μ–Czochralski process using excimer-laser and sputtered a-Ge layer. TFTs were fabricated inside the single grain of Ge. Capping silicon dioxide was applied before the laser crystallization with which a high quality Ge/insulator interface was formed during the laser annealing. Source/drain regions were formed by doped Si instead of the Ge, which ensures a low contact resistance and suppresses fast dopant diffusion in the channel. N- and p-channel SG Ge TFTs showed electron and hole mobility of as high as 3337cm^2/Vs and 1719cm^2/Vs, respectively. On/off current ratio for the both types of TFTs was in the order of 10^7.

Introduction

Germanium (Ge) has long been known to possess superior electron and hole mobilities than silicon (Si). [1] The large electron mobilities are highly desirable for high speed electronic circuits based on both ULSI and TFT technologies. In addition, Ge can absorb light with a longer wavelength because of the narrow band-gap. Therefore a near infrared photo-diode is expected to be realized for use as optical interconnect in future ULSI. Therefore both MOSFET and TFT using Ge have been recently studied extensively.

In the Ge MOSFET, there are several issues in the device fabrication point of view. First problem is the high interface states density between the Ge and an insulator which lowers carrier mobility [2]. Second problem is the fast dopant diffusion into channel during impurity annealing, which shortens the channel and increases the off-current.

In the TFTs using Ge channel, device has been fabricated with polycrystalline-Ge on a glass substrate. Because the film is polycrystalline, mobility becomes even lower by scattering at grain boundaries [3]. Off current becomes also very large due to carrier generation at the boundaries, resulting in very small on/off current ratio.

Those problems will be solved if the transistor can be made inside a single grain of Ge with a new device process realizing high quality Ge/gate insulator interface and negligible doping diffusion into the channel. Here, accurate 2D location control of large Ge grains is essential to ensure uniform property. For Si, we have reported that location of individual Si grain can be controlled at predetermined position by a developed process so called μ-Czochralski process [4,5] based on pulsed-laser crystallization. Grain size of the Si grains can be up to 9 μm and TFTs fabricated inside the single Si grain show high carrier mobilities for electron and holes (600cm^2/Vs and 250cm^2/Vs), despite of the low-temperature (<350°C) process [6]. We have also demonstrated RF circuit with low-noise amplifier operating at 440MHz [7] and monolithic 3D-IC with CMOS inverters with stacked two SG-TFTs [8]. If the process is applied to Ge, we expect even better electrical properties owing to the high carrier mobilities.

In this paper we will review our recent achievements in location-control of Ge grains and high performance single-grain (SG) Ge thin film transistor (TFT) fabricated inside the Ge grain. Large Ge grains having a maximum grain size of as large as 10 μm were obtained at predetermined positions using excimer-laser. A high quality Ge/insulator interface was formed with a capping silicon dioxide which was annealed during the laser crystallization. TFTs were fabricated inside the single Ge grains with source and drain made with doped-Si. N- and p-channel SG Ge TFTs showed electron and hole mobility of 3337cm^2/Vs and 1719cm^2/Vs, respectively. On/off current ratio for the both types of TFTs was in the order of 10^7.

LOCATION CONTROL OF GERMANIUM GRAINS

We first discuss 2D location control of germanium grains with the μ-Czochralski process based on excimer-laser crystallization.

The μ-Czochralski (grain-filter) process with germanium

We have proposed the μ-Czochralski process for 2D location control of semiconductor grains. As depicted in Fig. 1(a), the μ-Czochralski process has a locally increased thickness of the semiconductor film in a cavity (grain-filter) in a substrate. Upon excimer-laser irradiation, the semiconductor film surrounding the grain-filter melts completely, whereas the grain filter won't melt completely due to the large heat dissipation and heat capacitance there. During vertical growth of pre-existing seeds in the grain filter, occlusion of grains occurs reducing the number of growing grains. By increasing the aspect ratio of more than about 7, only single grain can be filtered out from the many pre-existing fine grains. In case of Si, we have demonstrated that, using a 250 nm thick LPCVD a-Si film as a precursor, a single grain of Si with a size of up to 9 μm can be positioned on top of the grain-filter. [4,5] We applied the method for 2D location control of Ge grains.

Experimental

2D location control of germanium grains are obtained as follows [9]. Firstly, 1 μm diameter holes are formed in oxide and 840 nm thick SiO$_2$ layer is then 2nd SiO$_2$ was deposited by TEOS-PECVD at a substrate temperature of 350 °C in order to reduce the grain-filter diameter to approximately 100 nm. An etch-back of the SiO$_2$ is needed to make slope of sidewall of the grain-filter gentle for the following sputtering deposition of 250 nm thick a-Ge film at different substrate temperatures of 100 °C, 400 °C, and 550 °C. Sputtering is performed with pulsed DC mode with a power of 1 kW, frequency of 75 kHz, pulse with of 496 ns and duty cycle of 3.7 percent.

Then a single pulse of XeCl (308nm, 25ns) excimer laser irradiates the Ge surface with energy densities ranging from 200 up to 800 mJ/cm². Figure 1(b) shows an example of SEM image of location-controlled Ge grain with a-Ge deposited at 550°C [9]. It can be seen that Ge grains having almost squared shape with a side length of 6 μm are successfully positioned on the grain filters. It can be also noticed that the surface inside the germanium grain is flat despite the gentle slope of the holes and hence initial indentation of Ge surface there. This is because the melting of the a-Ge by the laser pulse induced mass transfer into the grain filter and flattened the surface.

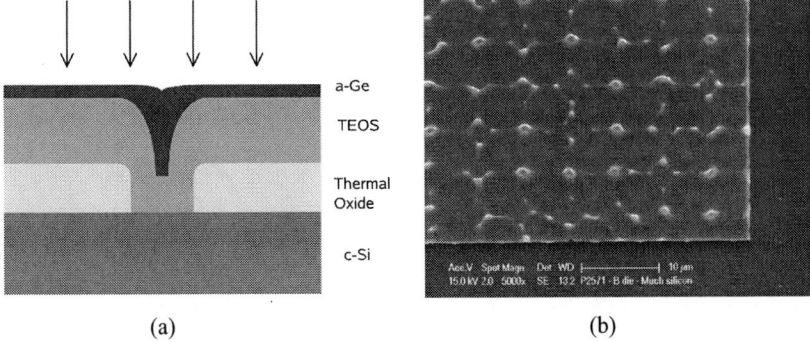

(a)   (b)

Figure 1. Schematic diagram (a) and SEM image (b) of location-controlled germanium grains by the μ-Czochralski process. Pitch of the grain-filters is 6 μm.

Grain size

Figure 3 shows grain size of the location-controlled Ge as a function of energy densities of the excimer-laser. In the figure, as a reference, grain size of Si deposited with LPCVD with the same thickness is also plotted. For the both Ge and Si, the grain sizes increase with the energy density. Difference between the two is however that the location control of Ge grains is obtained by much lower energy range than that of the Si. This is

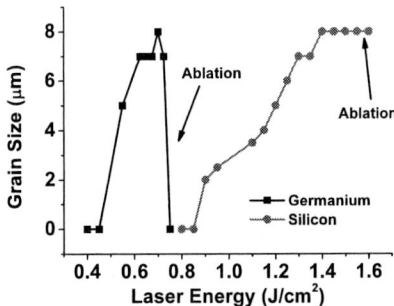

Figure 2. Grain size of location-controlled Ge (sputtered at 550 °C) and Si (LPCVD at 545 °C) versus laser energy density

consistent with the lower melting temperature of the Ge than that of the Si. Maximum grain size of Ge is about 7 μm, which is slightly smaller than that of Si (8 μm). After reaching the maximum grain size, it drops dramatically to the small value of about a few hundred nano meters. In this regime where the grain filter seems to be completely melted, many of those small grains are uniformly grown over the surface and there is no evident location-control of the grains observed. Such behavior is notably different than the case of Si, where still large grains are grown on the grain filter even after complete melting the grain filter and reaching the maximum grain size. [5] Although exact reason for this difference is still under investigation, it indicates clearly difference in nucleation behavior between the molten-Ge and molten-Si. For both materials, the film gets agglomerated with an excess energy density.

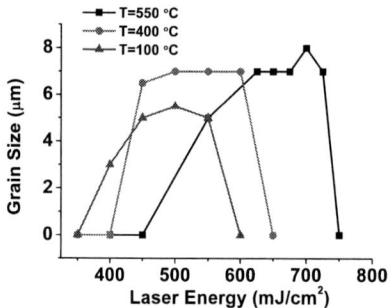

Figure 3. Grain size of location-controlled Ge for various deposition temperature

We have performed similar experiment with a-Ge sputtered at different temperature T. As shown in the Fig. 3, with decreasing the temperature of the sputtering, the onset of the grain size enlargement occurs at lower energy densities. For T of 400 °C, the process window becomes slightly smaller (450-600 mJ/cm²) and the grain size can reach 7 μm at most. For T of 100 °C, the maximum grain size decreased to 5 μm. After reaching the maximum grain size, the behavior is the same for all temperature; namely, the sudden drop of grain size and the random formation of the grains, followed by the ablation at an excess energy.

The maximum grain size was further increased by elongating the pulse duration. With a longer pulse duration of 180 ns, as shown in the Fig. 4 (a), the grain size can reach 10 μm, which is even larger than that of Si deposited by LPCVD (10 μm). The enlargement of grain by the longer pulse duration is explained by the increased amount of heat deposited directly under the germanium due to the long diffusion time and onset delay in homogeneous nucleation in the molten-Ge, which stops the lateral grain growth from the grain filter and eventually limits the maximum grain size.

Crystal quality and residual stress in the Ge grain were measured by micro-Raman spectroscopy. As shown in Fig. 4(b), the Raman shift peak of SG-Ge is at a lower wave number than that of Ge bulk by 3.812cm⁻¹ which suggests high tensile stress is formed in the SG-Ge during ELC because of the large difference in the heat expansion coefficient.

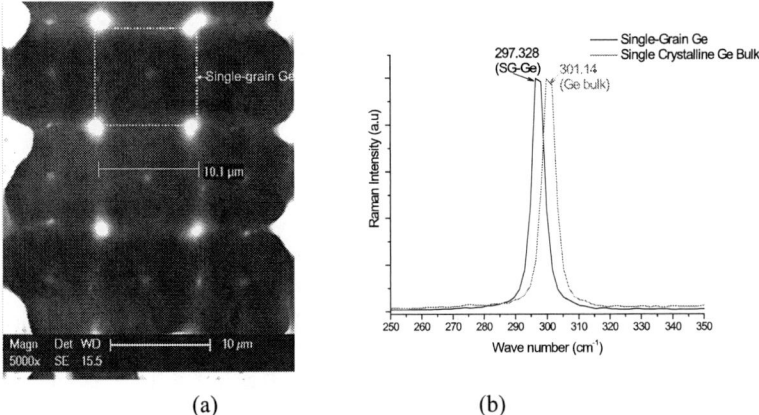

(a)   (b)

Figure 4. SEM (a) and Raman shift (b) of location-controlled Ge grains obtained with a longer pulse duration of 180ns.

Microstructure of germanium grains

We have analyzed microstructure of the obtained location-controlled Ge grain with the electron backscattering diffraction (EBSD). As shown from the Fig. 5(a), the planar defects generating from the grain filter holes are only coincident site lattice (CSL) with meanly $\Sigma 3$ and $\Sigma 9$, whereas random grain boundaries are only on the edge of the grain where others grains collide. Similar result has been obtained for Si as well. [10] Electrical activities of the CLSs must be negligible as in the case of Si. The quality of the single crystal of Ge depends on the grain filter diameter. In fact, as shown in Fig. 5(b), for the large diameter of 500 nm of the grain filter, random grain boundaries are incorporated in the location-controlled grains due to decreased filtering of grains during vertical growth in the grain filter. Similar trend has been observed for the case of Si. [10]

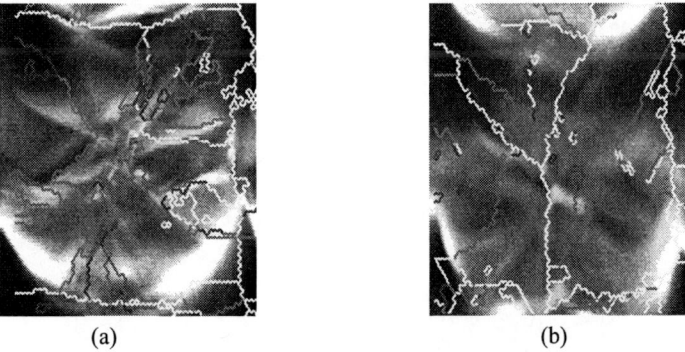

(a)   (b)

Figure 5. Electron backscattering diffraction (EBSD) using grain filter size of 100 nm (a) and 500 nm (b). The yellow lines are random grain boundaries, the red: $\Sigma 3$ CSL boundary and the blue: $\Sigma 9$ CSL boundaries.

Single-grain germanium TFTs

In this section we propose a new device structure for Ge FETs and review fabrication process and electrical properties of single-grain Ge TFT made inside a Ge grain.

Single-grain Ge TFTs

In the Ge MOSFET, device is fabricated in germanium on insulator (GOI) substrate because of advantage of low RF loss. Apart from difficulty in obtaining defect free GOI substrate, there are several issues in the device fabrication point of view. First problem is the high interface states density between the Ge and an insulator which degrade subthreshold swing and lowers carrier mobility. [2] The traps are created due to desorption of GeO [11-13], which is native germanium oxide and unstable chemically and thermally.

Second problem in the Ge transistor is about doping in the source and drain. Because of the much lower solubility limit of Ge than that of Si, contact resistance with a metal and parasitic resistance in the source and drain region become very high, resulting in low on-current. Also because of the same reason, the dopants diffuse easily into the channel, which leads to high leakage current. In addition, the dopants in Ge need to be annealed at higher temperature than that for Si to restore crystal damage due to implantation and interstitials made during the process.

To overcome those problems we propose a new device process and structure of Ge transistor. [16] As shown in Fig. 6, the device has a thin SiO_2 layer as gate insulator which is deposited at a low-temperature before laser crystallization of Ge. Upon irradiation of a pulsed-laser, the Ge surface absorb the light and heats up and eventually melts the film completely. (If there is grain-filter underneath, the grain-filter won't be melted.) The surface GeO is also completely melted, however it would not be desorbed and will be kept at the interface because of the presence of the capping SiO_2. After turning off the laser pulse, the molten-Ge starts solidifying. At the end of the solidification of Ge, we expect that an atomically thin, high quality GeO_2 layer [14,15] grows at the interface of Ge/SiO_2 by the presence molten GeO and oxygen atoms in the SiO_2. During the whole crystallization process of Ge, the SiO_2 is annealed by the heat and the bulk property is also improved.

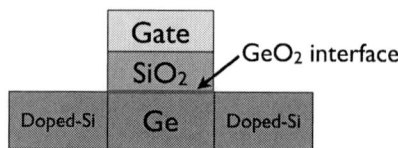

Figure 6. Proposed structure of the single-grain germanium TFTs.

The device is equipped with doped-Si for the source and drain regions. The dopants are implanted by an ion-implantation and activated by a pulsed-laser. The high solubility of Si ensures a low contact resistance to a metal and also a low source and drain parasitic resistance. The high solubility also prevents dopant diffusion into the channel. The pulsed-laser annealing effectively activate the dopants in a very short time without causing unwanted diffusion of dopants into the channel region.

Fabrication process

Figure. 7 (a)-(f) show schematic cross section of fabricated device [16]. Location-control of Ge grains is performed basically the same procedures as the aforementioned process. In this process, before depositing a-Ge into the 100nm diameter grain-filter, in order to control the crystal orientation of germanium, we first prepared (001) oriented Si seed inside the grain filter hole to control the orientation of the Ge grains. The textured seed was fabricated by MILC (metal induced lateral crystallization) which has been reported in our earlier paper [16] (Fig. 7(a)-(b)). CMP polished a way the Si film leaving only the seed of (001) oriented Si inside the grain filters. Then we have deposited a 250nm thick a-Ge layer by sputtering at a substrate temperature of 450°C. Then a 50nm thick SiO$_2$ layer was deposited by TEOS-PECVD at 350°C, as a capping layer for the following excimer laser crystallization (Fig.7(c)). The sample was crystallized by excimer-laser (λ=308nm, t=180ns) with a substrate temperature of 450°C (Fig.7(d)). During the laser crystallization, Ge grains grew epitaxially from the (001) oriented Si seed (Fig. 7(c)). After patterning the Ge and the capping SiO$_2$ into islands, additional 40nm SiO$_2$ was deposited by the TEOS PECVD. The total thickness of the gate oxide is

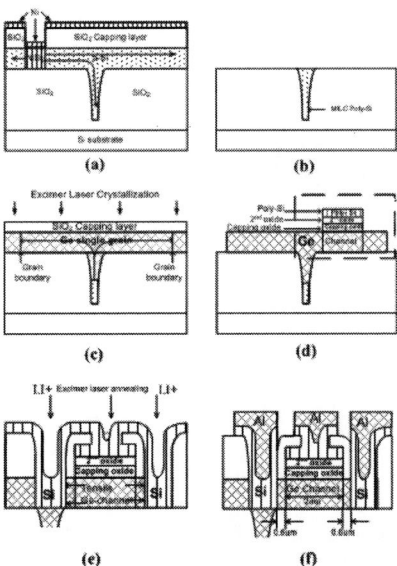

Figure 7. Schematics of fabrication process of the single-grain Ge TFTs

90nm. Then a 250nm thick a-Si was deposited as poly-gate by LPCVD at 545°C and phosphorous and boron ions were implanted with 10^{16} ions/cm^2 for n- and p-channel TFTs, respectively (Fig.7(d)). Then the dopants were activated by excimer laser annealing (ELA) with energy density of 300mJ/cm^2 at room temperature. The poly-Si and the gate SiO$_2$ were etched with a gate mask. After deposition of a passivation SiO$_2$, contact hole for source and drain was opened. The Ge in the source/drain contact hole was etched away by anisotropic plasma etching, leaving a 0.6 μm offset of intrinsic Ge adjacent to the channel (Fig.7(e)). A 250nm thick a-Si was deposited as source/drain by LPCVD and the same implantation and annealing scheme was repeated (Fig.7(e)). The channel width and length was 2μm for both p- and n-channel TFTs. After deposition and patterning Al pads, the sample was annealed at 400°C in H$_2$ (Fig.7(f)).

TFT characterization

Figure 8 shows transfer characteristics of the single-grain (SG) Ge TFTs. Effective mobility was calculated by transconductance (g_m) at Vth. For n-channel, the SG Ge TFT shows an effective mobility of 3337cm^2/Vs with a large on/off ratio of 10^8 at V_{DS}=0.1V. For p-channel, SG Ge TFT shows an effective mobility of 1719 cm^2/Vs with a large on/off ratio of 10^7 at V_{DS}=0.05V. The S value for n- and p-channel SG TFT is 0.19/dec. and 0.17V/dec., respectively. The high mobility and low S value suggest that a good interface property between Ge and gate SiO$_2$ was achieved due to the thin GeO$_2$ interface layer. The off-current was 10pA and 0.1pA for n- and p-channel, respectively, which suggest that phosphorous diffusion into Ge is effectively suppressed by the Ge offset and the pulsed-laser activation.

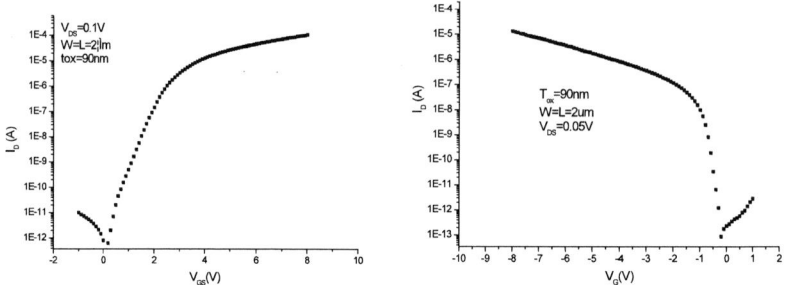

Figure 8. Transfer characteristics of n- (a) and p-channel (b) single-grain germanium TFTs

Figure 9 shows output characteristic of n- and p-channel devices. For the both types of devices, the current increases exponentially. Similar property has been observed for the tunneling MOSFETs. In addition the current does not saturate in, at least, the range of the V_{DS}. Those trends are explained in terms of band diagrams as shown in Fig. 10. At a low drain bias V_{DS}, with increasing the gate bias V_{GS}, the source barrier between Ge channel and doped-Si becomes so low that the electrons can emit through the intrinsic Ge barrier into the channel (Fig.10 (b)). The electron flow is however hindered by potential barrier at the Ge offset at drain side. By increasing the V_{GS} further the electrons are collected at the drain region due to the lowered barrier at the Ge offset at drain (Fig. 10(c)). With increasing the V_{DS}, the barriers at the intrinsic Ge offset at source and drain become so low that a lot of electrons can emit into the channel from the source and collected by the drain which result in very high drain current (Fig.10 (e)).

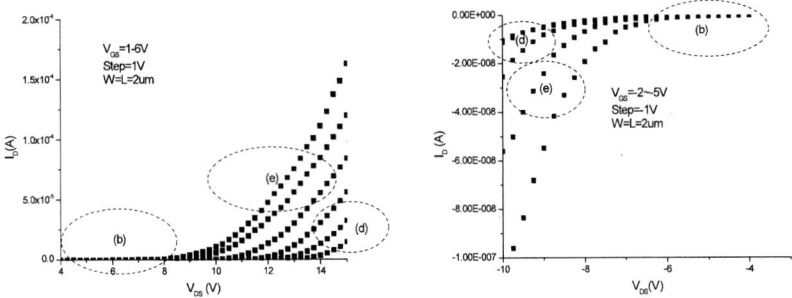

Figure 9. Output characteristics of n- (a) and p-channel (b) single-grain germanium TFTs

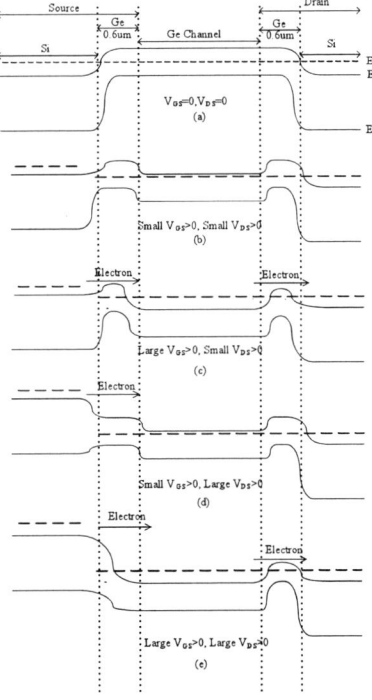

Figure 10. Band diagram of the single-grain Ge TFTs explaining possible transport mechanism of the TFTs with bias conditions of $V_{GS}=V_{DS}=0$ (a), low V_{DS} with low (b) and high V_{GS} (c) and high V_{DS} with low (d) and high V_{GS} (e).

Conclusions

We reviewed our recent achievements in location-control of Ge grains and high performance single-grain (SG) Ge thin film transistor (TFT) fabricated inside the Ge grain. Large Ge grains having a grain size of 10 μm were obtained at predetermined positions by the μ–Czochralski process using excimer-laser and sputtered Ge layer. TFTs were fabricated inside the single grain of Ge. Capping silicon dioxide was applied before the laser crystallization with which a high quality Ge/insulator interface was formed during the laser crystallization process. Source and drain were formed by doped Si instead of the Ge counterpart, which ensure a low contact resistance and suppresses fast dopant diffusion in the channel. N- and p-channel SG Ge TFTs showed electron and hole mobility of as high as 3337cm^2/Vs and 1719cm^2/Vs, respectively. On/off current ratio for the both types of TFTs was in the order of 10^7. This will open several new applications in TFTs for displays with ultra-wide-band RF wireless communication interface and monolithic 3D-ICs with optical interconnect.

Acknowledgments

The authors would like to thank the Dutch Technology Foundation STW for their financial support and the cleanroom staff of DIMES technology center for their support during the fabrication of these devices.

References

1. R. H. Reuss, et al., *Proceeding of IEEE,* vol. 93, no. 7, (2005)
2. K. Saraswat, et al., *IEDM* 2006, 659.
3. T. Sadoh, et al., *Jpn. J. Appl. Phy.,* 46, 3B, (2007) 1250
4. R. Ishihara, et al., *Proc. SPIE,* **4295**, 14 (2001)
5. R. Ishihara, et al., Solid State Electronics, **52**, 353-358 (2008)
6. V. Rana, et al., *IEEE Trans. Electron Devices*, **52**, 12, 2622 (2005)
7. N. Saputra, et al., *J. Solid State Circuits*, **43**, 7, 1563-1576 (2008)
8. M.R.T. Mofrad, et al., *Japanese Journal of Applied Physics*, vol. 48, (2009)
9. A. Baiano et al., *Trans. ECS*, 16(9), 153, (2008).
10. R. Ishihara, et al., *J. Crystal Growth*, **299**, 2, 316 (2007)
11. E.P. Gusev, et al., *Appl. Phy. Lett.,* 85, 2334. (2004)
12. C.O. Chui, et al., *IEEE Electron Device Lett.,* 23, 473 (2002)
13. D.S.Yu, et al., *IEEE Electron Device Lett.,* 25, 559 (2004)
14. D. Kuzum, et al., *IEEE Electron Device Lett.,* 29 (4) (2008) 328-330
15. C.H.Lee, et al., *Appl. Phys. Express* 2 (2009) 071404
16. T. Chen, et al., *IEEE Trans. Electron Devices,* 58, p. 223 (2010)

CHAPTER 2

CHALLENGES IN OXIDE AND ORGANIC TFTS

Success in Measurement the Lowest Off-state Current of Transistor in the World

Y. Sekine, K.Furutani, Y. Shionoiri, K. Kato, J. Koyama and S. Yamazaki

Semiconductor Energy Laboratory Co., Ltd.,
398 Hase, Atsugi, Kanagawa, 243-0036, Japan

An oxide semiconductor has a wide band gap and low off-state current. We focused on the off-state characteristics of a transistor including an oxide semiconductor and concentrated on measuring the off-state current. We fabricated a circuit TEG that had an In-Ga-Zn-Oxide TFT (IGZO-TFT) with a large channel width of 100,000 μm over a glass substrate and estimated the amount of electric charge flowing through the circuit TEG for a long time. As a result, we discovered that an IGZO-TFT had off-state current as small as 50yA/μm (at 85°C). We are the first to measure off-state current on the order of yA; "y" is the SI prefix representing 10^{-24}.

Introduction

In recent years, research and development for applications and devices using oxide semiconductors has been actively performed (1-5). In particular, the oxide semiconductors have attracted attention as a main material of next-generation displays, and we have been developed applications and devices using oxide semiconductors (6-17). This is because a transistor including an oxide semiconductor has higher on-state current, higher mobility, and lower off-state current than an a-Si transistor. We have reported that In-Ga-Zn-Oxide (IGZO) that was one of oxide semiconductor materials had a wide band gap (Eg = 3.15 eV) and a transistor including IGZO had low off-state current (23). Low off-state current contributes to decrease in power consumption. In addition, in the case where an IGZO transistor is used as a switching device, data retention is expected to be good. The IGZO transistor is expected to be applied to displays of personal computers and personal digital assistants that mainly display still images, for example, and devices that are specialized to store data, such as memory.

We reported that the frame rate was reduced in displaying still images by applying the low off-state current of the IGZO-TFT for displays, so that we decreased power consumption of the displays (18-23). Further, in the field of ULSI, we apply the IGZO-TFT to non-volatile memory (NVM) that has high endurance and is driven at high speed (24). However, the precise value of the off-state current of the IGZO-TFT is not clear. By knowing the precise value of the off-state current of the IGZO-TFT, the range of application of the IGZO-TFT is expected to spread.

This time, we concentrated on measuring the off-state current of the IGZO-TFT. We report that we measured extremely low off-state current of the IGZO-TFT by using a novel measurement method, a proper structure of the transistor, a proper evaluation system, and proper environment for evaluation. In addition, we also discuss application of the IGZO-TFT to memory.

Electrical Characteristics and Stacked Structure

Fig. 1 shows the I_D-V_G characteristics of an IGZO-TFT. Fig. 2 shows a photomicrograph of the IGZO-TFT. The transistor has a channel length L of 3 μm and a channel width W as large as 1 m. The large channel width makes off-state current high and facilitates detection of the off-state current. Note that a semiconductor parameter analyzer was used for the measurement. From Fig. 1, it is found that the off-state current is lower than the lower detection limit (1×10^{-13} A) of the measurement system even when the IGZO-TFT has a channel width as large as 1 m. Accordingly, the precise value of the off-state current is even lower.

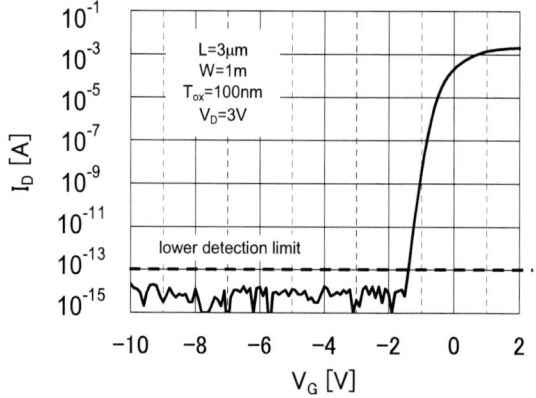

Figure 1. I_D-V_G Characteristics of IGZO-TFT

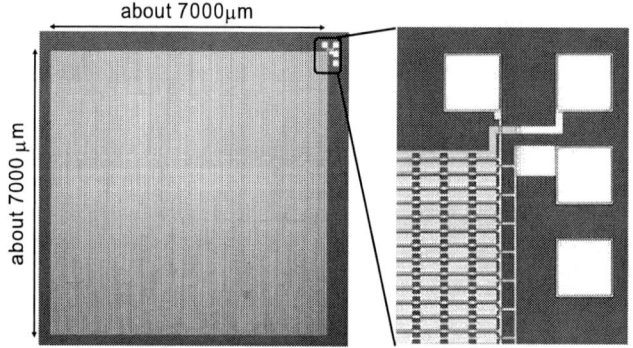

Figure 2. Photomicrograph of IGZO-TFT with Channel Width W of 1 m

Fig. 3 shows a cross-sectional structure of our IGZO-TFT. The IGZO-TFT is a top gate transistor. In addition, the IGZO-TFT has an overlap structure in which a gate electrode overlaps with source/drain electrodes. The overlap width is 2 μm. The IGZO-TFT is formed by stacking the following over a glass substrate: a base film, an active layer, source/drain electrodes, a gate insulator, a gate electrode, and a passivation layer.

The base film is a 300 nm silicon oxide film. The active layer is a 30 nm IGZO film. The source/drain electrodes are a 100 nm tungsten film. The gate insulator is a 100 nm silicon oxide film. The gate electrode is a stack of a 15-nm-thick tantalum nitride film and a 135-nm-thick tungsten film. The passivation layer is a 300 nm silicon oxide film.

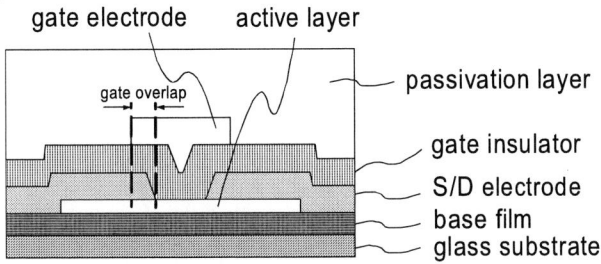

Figure 3. Schematic Cross-Sectional View of IGZO-TFT with Overlap Structure

Experiment

Basic Concept

By the conventional measurement method, the off-state current of the IGZO-TFT is found to be lower than the lower detection limit of the measurement system (1×10^{-13} A). We tried measuring the off-state current of the IGZO-TFT, which was expected to be very low, by our unique measurement method. A measurement method we developed is a method in which the amount of electric charge flowing through the IGZO-TFT is measured for a long time so as to estimate the amount of current.

Fig. 4(a) shows a conceptual diagram of a circuit for measuring off-state current. Fig. 4(b) shows change in the voltage at a node F over time. The node F is connected to an IGZO-TFT (DUT) to be evaluated and load capacitance C. The off-state current I of the DUT is expressed by the following equation [1].

$$I = C \, \Delta V / \Delta t \qquad [1]$$

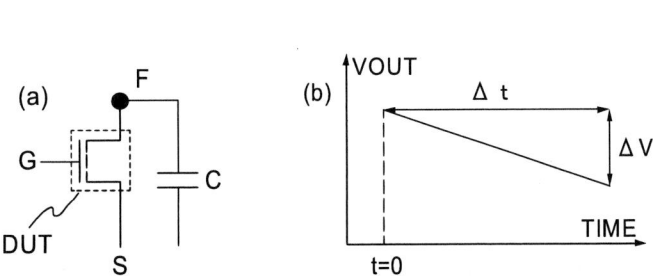

Figure 4(a). Conceptual Diagram of Circuit for Measurement (b). Change in Voltage at Node F Over time

By measuring the change ($\Delta V/\Delta t$) in the voltage of the node F over time, the off-state current I can be estimated. In particular, when time for measurement becomes longer, lower off-state current can be measured. Therefore, the measurement method is effective for measuring extremely low current.

Circuit Configuration and Stacked Structure

Fig. 5 shows a block diagram of a circuit TEG formed over a glass substrate. The circuit TEG includes the DUT, a programming circuit for writing data to the node F, and a reading circuit for reading the voltage of the node F.

The channel width of the DUT is as large as 100,000 μm. The large channel width makes off-state current high and facilitates detection of the off-state current.

The channel width of an IGZO-TFT included in the programming circuit is 10 μm, which is sufficiently smaller than that of the DUT (i.e., 1/10,000); thus, the accuracy of detecting the off-state current of the DUT is prevented from being decreased.

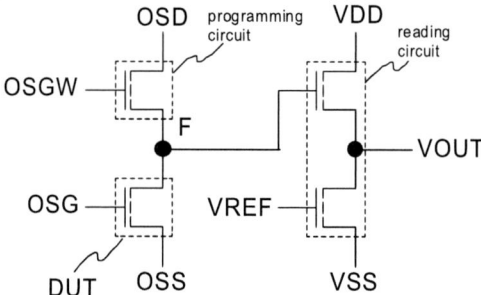

Figure 5. Block Diagram of Circuit TEG

Further, according to the equation [1], decrease in the load capacitance C as well as increase in time for measurement is effective for measuring the extremely low current of the IGZO-TFT.

We tried decreasing the load capacitance C. First, a source follower circuit is used as the reading circuit. Second, an offset structure is employed for the DUT. Third, the IGZO-TFTs are fabricated over a glass substrate.

A source follower circuit used as the reading circuit has a feature of small input capacitance. Therefore, load capacitance can be reduced, so that the off-state current can be estimated with high accuracy. Fig. 6 shows input-output characteristics of the source follower circuit ($n = 3$). In the experiment, Vref is 0.5 V and Vi is 3 V. In Fig. 6, the slopes (dVo/dVi) in the case of Vi = 2 V to 4 V are 0.912, 0.923, and 0.915. This shows good linearity with small variation.

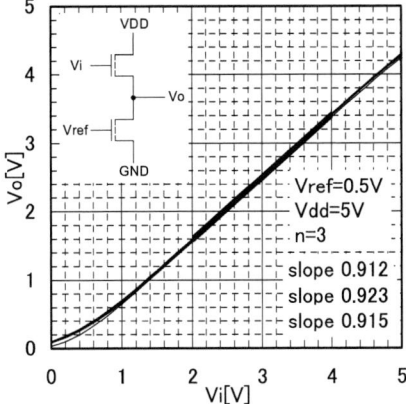

Figure 6. Input-Output Characteristics of Source Follower Circuit

An offset structure shown in Fig. 7 is employed for the IGZO-TFTs. In the overlap structure shown in Fig. 3, the gate electrode of the DUT overlaps with its drain electrode; therefore, the load capacitance C becomes large. As a result, the measurement accuracy for evaluating low off-state current is decreased. When the IGZO-TFT has the offset structure instead of the overlap structure, so that the load capacitance per channel width of 1 μm can be approximately 1/14, from 1 fF/μm to 0.07 fF/μm. By decrease in load capacitance, the off-state current can be estimated with high accuracy.

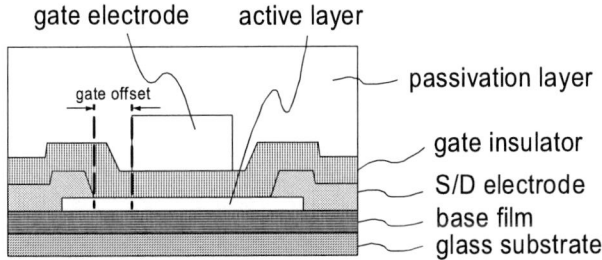

Figure 7. Schematic Cross-Sectional View of IGZO-TFT with Offset Structure

In order to reduce the capacitance between the substrate and the node F, the IGZO-TFT is fabricated over a glass substrate instead of a Si wafer. The load capacitance C connected to the node F includes the capacitance between the node F and the substrate, in addition to the capacitance between the gate electrode and the drain electrode of the DUT, the capacitance between the gate electrode and the drain electrode of the IGZO-TFT included in the programming circuit, and the capacitance of the gate electrode of a transistor included in the reading circuit. Table I shows the result of evaluating each capacitance. The capacitance between the node F and the glass substrate provided therebetween is as small as 3.7 % (0.28 pF) and thus almost negligible.

Table I. Components of Load Capacitance C of Node F

Components	Value	Proportion
Capacitance between gate electrode and drain electrode of IGZO-TFT to be evaluated	6.9×10^{-12} F	91.2 %
Capacitance between gate electrode and drain electrode of IGZO-TFT included in programming circuit	7.3×10^{-15} F	0.1 %
Gate electrode of transistor included in reading circuit	3.2×10^{-13} F	4.36 %
Capacitance between node F and substrate	2.8×10^{-13} F	3.7 %

Measurement

In order to surely turn off the DUT, the voltage (OSG) applied to the gate electrode in Fig. 5 was −3 V and the voltage (OSS) applied to the source electrode was 0 V. Fig. 8 shows a timing chart and the voltage of each terminal. The voltage of the node F was set to 3 V. As programming operation, OSGW of 5 V was applied to the gate electrode of the IGZO-TFT included in the programming circuit for 10 seconds. OSGW of 5 V is a sufficient level for applying 3 V to the node F. Ten seconds are sufficient time for applying 3 V to the node F. In addition, we confirmed that the voltage of the node F in the first programming operation is at the same level as that in the second programming operation. Further, in the period other than the programming operation, OSGW was set to −3 V to surely turn off the DUT. As reading operation, the reading circuit was activated for 10 seconds every 5 minutes. In the reading operation, the output voltage VOUT in Fig. 5 was read. In the period other than the reading operation, the reading circuit was inactivated to reduce the BT stress applied to the reading circuit. In the measurement, one set is determined as follows: the programming operation is performed once and then reading operation is performed plural times (e.g., 144 times within approximately 12 hours). In order to examine reproducibility, the set was repeated plural times.

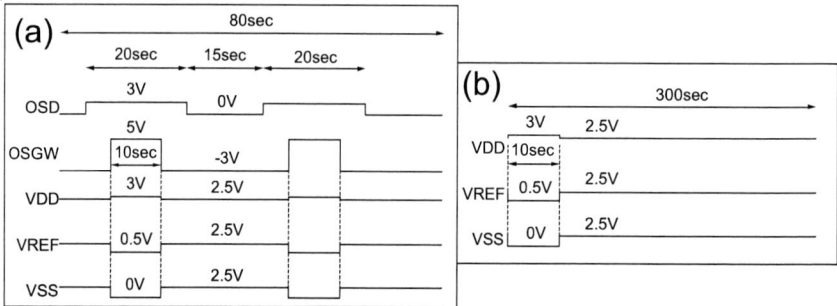

Figure 8(a). Timing Chart of Programming Operation (b). Timing Chart of Reading Operation

Measurement environment is explained. According to the equation [1], it is also important to detect slight voltage change ΔV with high accuracy in order to measure the extremely low current of the IGZO-TFT. Thus, it is important to adjust measurement environment in order to suppress measurement noise at the node F and perform measurement stably for a long time. In particular, the current characteristics of the IGZO-TFT are influenced by temperature, humidity, and light; therefore, temperature, humidity, and light were kept constant. Specifically, measurement was performed at 85 °C, in dry air (having a dew point of lower than −60 °C), and in the darkness. We also paid attention to a power supply system. We provided a vibration removal board under the stage for measuring the TEG. As a result, variation in measurement of the voltage of the node F is suppressed to 1 mV or lower for as long as 12 hours or more.

Result

Fig. 9 shows the result of the measurement performed at 85 °C ($n = 2$). The measurement condition is as follows. The DUT had a channel length L of 3 μm, a channel width W of 100,000 μm, and the offset structure (with a width of 2 μm). In the measurement, one set is determined as follows: programming operation is performed once and then the reading operation was performed 144 times within approximately 12 hours. The set was repeated at least four times.

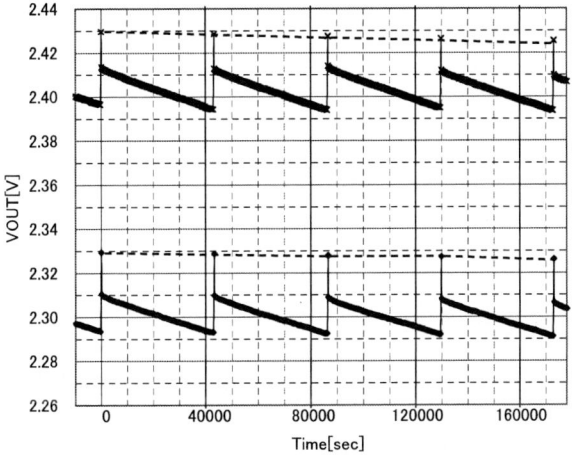

Figure 9. Change in Output Voltage VOUT Over Time

The off-state current of the DUT was estimated on the basis of the equation [1]. Data in the last 6 hours of each set was used in the estimation, and the slope of the output voltage VOUT was estimated by a least squares method. In the programming operation performed every 12 hours, as shown by a dotted line in Fig. 9, the output voltage VOUT varied slightly although a voltage of 3 V was written to the node F. We considered that the variation was caused because the characteristics of the reading circuit varied as time passed. Accordingly, the off-state current was corrected by subtracting the slope of the dotted line from the formerly-estimated slope of the output voltage VOUT. Assuming that all the estimated off-state current flows through the DUT, the current value per 1-μm channel width of the IGZO-TFT is calculated. Table II shows the result of the estimation after the correction of the off-state current. As a result, the off-state current of the IGZO-

TFT at 85 °C was estimated as approximately 50 yA/μm (n = 2). Variation among samples and sets were small.

Table II. Estimation after Correction of Off-state current

Time [sec]	21600-43200	64800-86400	108000-129600	151200-172800
Off-state current I of Sample 1 [yA/μm]	46	46	45	44
Off-state current I of Sample 2 [yA/μm]	43	41	41	40

Evaluation and analysis are performed at each measurement temperature of 125 °C and 150 °C, and the off-state current is estimated. Fig. 10 shows an Arrhenius plot. From Fig. 10, we confirmed that the off-state current had a good linearity at 85 °C, 125 °C, and 150 °C at which the measurement was performed. We estimated the off-state current at 27 °C by extrapolation on the basis of the good linearity in the Arrhenius plot. The off-state current was estimated to be approximately 3×10^{-26} A/μm.

In such a manner, we proved that the extremely low off-state current of the fabricated IGZO-TFT, which could not be measured by conventional methods, can be measured in the case where a proper measurement method is used, proper measurement environment is prepared, and the fabricated IGZO-TFT has a proper structure.

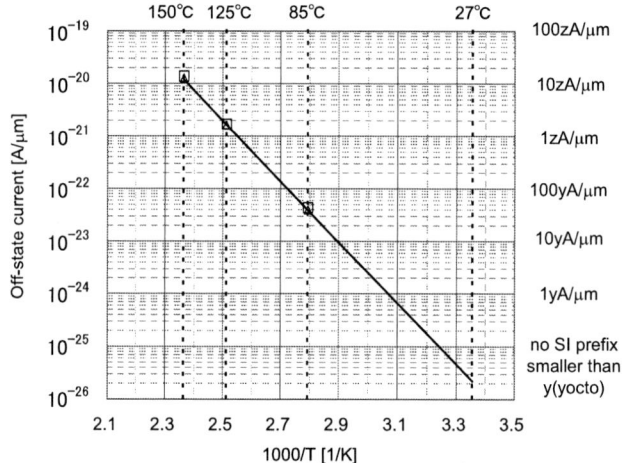

Figure 10. Arrhenius Plot

Application to Memory

The result of the measurement shows that the off-state current of the IGZO-TFT was approximately 50 yA/μm (at 85°C). This means that a switching device including the IGZO-TFT has extremely high charge retention and can be applied to NVM. On the basis of this result, we are developing Non-volatile Oxide Semiconductor Random Access Memory (NOSRAM) that is novel NVM (24). The distinguishing feature of NOSRAM is endurance much higher than that of conventional flash memory. Figs. 11(a) and 11(b) show a circuit configuration of a memory cell of NOSRAM and basic characteristics of the memory cell. The memory cell includes one IGZO-TFT, one Si transistor, and one capacitor. The capacitor can hold accumulated electric charge for a long time because of extremely low off-state current of the IGZO-TFT.

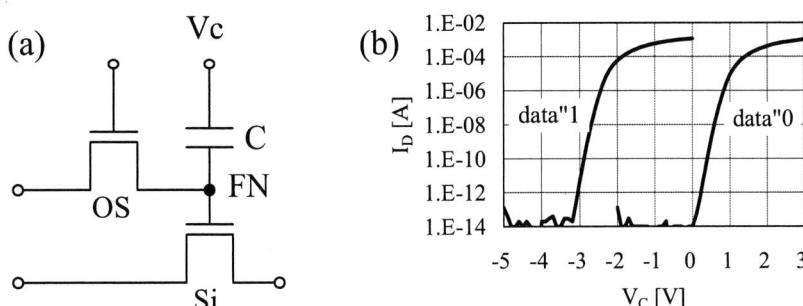

(a) Circuit Configuration of Memory Cell (b) Characteristics of Memory Cell
Figure 11. Memory Cell and Basic Characteristics

Fig. 12 shows data of measuring the endurance of the memory cell. We confirmed from Fig. 12 that Vth window width did not change even after data was rewritten 10^{12} times. As shown here, the NOSRAM has extremely high endurance.

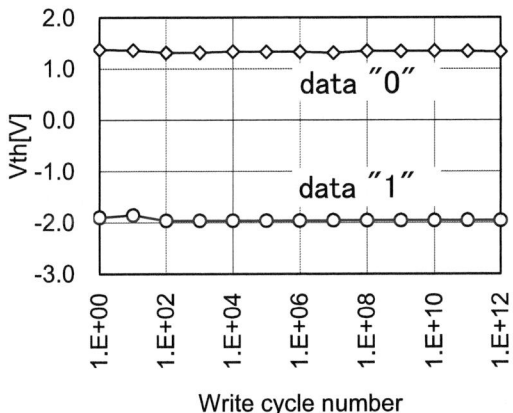

Figure 12. Memory Cell and Basic Characteristics

Further, we fabricated 1 Mbit NOSRAM and verified its operation. Fig. 13(a) shows photomicrographs of a 1 Mbit NOR chip and a NOR-type memory cell. Fig. 13(b) shows photomicrographs of a 1 Mbit NAND chip and a NAND-type memory cell. In the NAND-type chip, eight memory cells were connected in series. Fig. 14 shows data of measuring data retention of the prototyped chip of 1 Mbit NOR-type NOSRAM. We can confirm from Fig. 14 that the retention of data "1" in an acceleration test at 125 °C hardly changes after 100 hours or more. The acceleration factor at 125 °C with respect to room temperature is 10^4 or more (as expected from Fig. 10); therefore, data is expected to be retained for 10 years or more at room temperature.

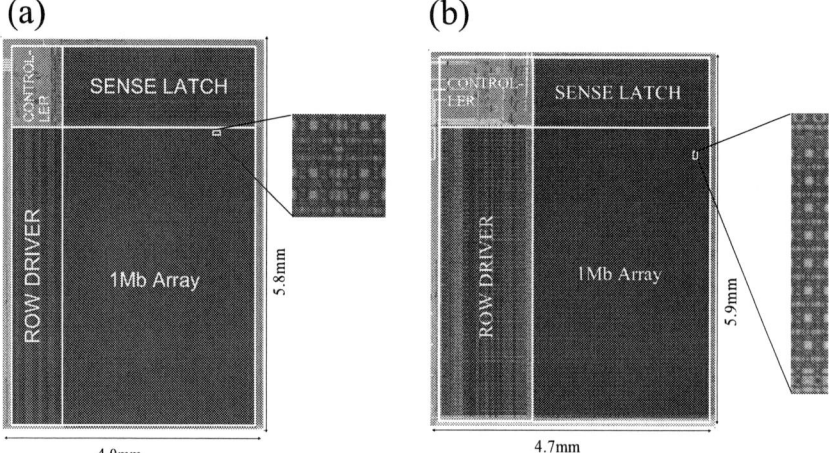

(a) 1 Mbit NOR chip and Memory cell (b) 1 Mbit NAND chip and Memory cell

Figure 13. Photomicrograph

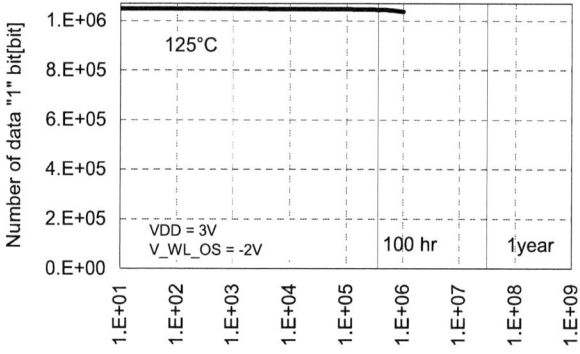

Figure 14. Data Retention (1Mbit NOR-type NOSRAM)

We confirmed that the fabricated NOSRAM not only had high endurance but also was able to rewrite data with 4.5 V or lower. That is to say, NOSRAM has a power consumption advantage over flash memory which needs 10 V or higher for rewriting data. In our view, NOSRAM is promising as memory for portable electronic devices. We are going to advance development of NOSRAM in the future.

The off-state current of the IGZO-TFT was measured as 50 yA/μm (at 85 °C) as described above, which proves that the IGZO-TFT can be applied to NVM.

Summary

We developed a method for estimating off-state current of the IGZO-TFT that is one of oxide semiconductors. We fabricated a circuit TEG including an IGZO-TFT with a large channel width and measured the amount of electric charge flowing through the circuit TEG. By the evaluation and analysis, we found that the IGZO-TFT had the off-state current as extremely low as approximately 50 yA/μm (at 85 °C), which mean we succeeded in measuring off-state current on the order of yA. The off-state current at 27 °C is estimated to be approximately 3×10^{-26} A/μm (extrapolation value). Moreover, we confirmed that the switching device including the IGZO-TFT is promising for being applied to NVM. We are further improving measurement accuracy and developing application of the IGZO-TFT to NVM.

References

1. N. Kimizuka and T. Mohri, *J. Solid State Chem.*, **60**, 382 (1985).
2. N. Kimizuka, M. Isobe, and M. Nakamura, *J. Solid State Chem.*, **116**, 170 (1995).
3. M. Nakamura and N. Kimizuka, T. Mohri, *J. Solid State Chem.*, **93**, 298 (1991).
4. N. Kimizuka and T. Mohri, Japanese Patents 1606567 (1991), 1639406 (1992), 1639407 (1992), 1606568 (1991), 1639398 (1992), and 1606571 (1991).
5. C. Li, Y. Bando, M. Nakamura, M. Onoda, and N. Kimizuka, *J. Solid State Chem.*, **139**, 347 (1998).
6. K. Nomura, H. Ohta, K. Ueda, T. Kamiya, M. Hirano, and H. Hosono, *SCIENCE*, **300**, 1269 (2003).
7. K. Nomura, H. Ohta, A. Takagi, T. Kamiya, M. Hirano, and H. Hosono, *Nature*, **432**, 488 (2004).
8. S. Maekawa and H. Kuwabara, Korean Patent, 1022352 (2011).
9. K. Akimoto, T. Honda, and N. Sone, Chinese Patent, 654582 (2010), 668939 (2010).
10. K. Akimoto, T. Honda, and N. Sone, US Patent, 7674650 (2010), 7732819 (2010), 7910490 (2011), 7932521 (2011).
11. K. Akimoto, Chinese Patent, 731689 (2011).
12. H. Suzawa, S. Sasagawa, and T. Muraoka, US Patent, 7915075 (2011).
13. T. Osada, K. Akimoto, T. Sato, M. Ikeda, M. Tsubuku, J. Sakata, J. Koyama, T. Serikawa, and S. Yamazaki, *Jpn. J. Appl. Phys.*, **49**, 03CC02-1 (2010).
14. H. Godo, D. Kawae, S. Yoshitomi, T. Sasaki, S. Ito, H. Ohara, H. Kishida, M. Takahashi, A. Miyanaga, and S. Yamazaki, *Jpn. J. Appl. Phys.*, **43**, 03CB04-1 (2010).
15. H. Ohara, T. Sasaki, K. Noda, S. Ito, M. Sasaki, Y. Endo, S. Yoshitomi, J. Sakata, T. Serikawa, and S. Yamazaki, *Jpn. J. Appl. Phys.*, **43**, 03CD02-1 (2010).

16. M. Takahashi, H. Kishida, A. Miyanaga, and S. Yamazaki, *IDW'08 Digest*, 1637 (2008).

17. J. Sakata, H. Ohara, M. Sasaki, T. Osada, H. Miyake, H. Shishido, J. Koyama, Y. Oikawa, H. Maruyama, M. Sakakura, T. Serikawa, and S. Yamazaki, *IDW'09 Proc.*, 689 (2009).

18. S. Amano, H. Harada, K. Akimoto, J. Sakata, T. Nishi, K. Moriya, K. Wakimoto, J. Koyama, S. Yamazaki, Y. Oikawa, T. Ikeyama, and M. Sakakura, *SID Symposium Digest*, **41**, 626 (2010).

19. H. Shishido, K. Toyotaka, M. Tsubuku, K. Noda, H. Ohara, T. Nishi, K. Moriya, H. Godo, J. Koyama, S. Yamazaki, Y. Oikawa, T. Handa, and M. Sakakura, *SID Symposium Digest*, **41**, 1128 (2010).

20. T. Nishi, K. Moriya, S. Fukai, Y. Kubota, K. Akimoto, J. Sakata, K. Kusunoki, R. Arasawa, K. Wakimoto, J. Koyama, S. Yamazaki, Y. Oikawa, K. Okazaki, and M. Sakakura, *SID Symposium Digest*, **41**, 1685 (2010).

21. H. Harada, A. Umezaki, S. Amano, T. Nishi, K. Moriya, K. Wakimoto, J. Koyama, S. Yamazaki, Y. Oikawa, T. Ikeyama, and M. Sakakura, *Proc. AM-FPD'10 Digest*, 199 (2010).

22. K. Toyotaka, K. Kusunoki, T. Nishi, K. Moriya, S. Fukai, Y. Kubota1, K. Wakimoto, J. Koyama, S. Yamazaki, Y. Oikawa, K. Okazaki, and M. Sakakura, *Proc. AM-FPD'10 Digest*, 17 (2010).

23. H. Godo, A. Miyanaga, K. Kusunoki, K. Toyotaka, T. Nishi, K. Moriya, S. Fukai, Y. Kubota, K. Wakimoto, J. Koyama, S. Yamazaki, Y. Oikawa, K. Okazaki, and M. Sakakura, *IDW'10 Proc.*, 235 (2010).

24. T. Matsuzaki, H. Inoue, S. Nagatsuka, Y. Okazaki, T. Sasaki, K. Noda, Y. Sekine, D. Matsubayashi, T. Ishizu, T. Onuki, A. Isobe, Y. Shionoiri, K. Kato, J. Koyama, Y. Yamashita, and S. Yamazaki, to be published in *International Memory Workshop* (2011).

Low Power 6.0-Inch Extended Graphics Array Reflective Liquid Crystal Display using Indium Gallium Zinc Oxide Semiconductor with Electronic Paper Function

Makoto Kaneyasu[a], Hiroyuki Miyake[a], Takeshi Nishi[a], Yoshiharu Hirakata[a], Jun Koyama[a], Rai Sato[b], Masayuki Sakakura[b], and Shunpei Yamazaki[a],

[a] Semiconductor Energy Laboratory Co., Ltd., 398 Hase, Atsugi, Kanagawa, 243-0036, Japan
[b] Advanced Film Device Inc., 161-2 Masuzuka, Tsuga-machi, Tochigi, Tochigi, 328-0114, Japan

> We focused on the ultra-low leakage current of indium gallium zinc oxide (IGZO) field-effect transistors (FETs). The power consumption of an liquid crystal display (LCD) panel can be markedly reduced by significantly lowering frame frequency, i.e., stopping unnecessary data rewrites at the time of displaying still images. We successfully fabricated a 6.0-in. extended graphics array (XGA) reflective liquid crystal display (RLCD) panel integrally including a data selection demultiplexer and a scan driver and capable of displaying still images at 1/180 fps. Thus, the panel is suitable for electronic paper displays.

Introduction

Electronic paper displays have become available on the market as novel thin display devices. In electronic paper displays, electrophoresis elements have memory functionality and the driving number of drivers can be significantly reduced. Thus, power consumption can be markedly reduced. Further, electrophoretic displays have the following advantages: texture similar to that of paper, high visibility, and applicability to flexible displays. Accordingly, electrophoretic displays have been widely used.

However, the electronic paper displays have problems as follows: high drive voltage of drivers needed for rewriting data, insufficiency of colorization and expression of gray levels, long rewrite time due to low response speed, and difficulty of displaying moving images. In particular, the low response speed is a major problem because operability is impaired.

In recent years, displays using oxide semiconductors have been actively developed. Concerning indium gallium zinc oxide (IGZO)-based oxide materials, Kimizuka et al. found the general formula of homologous, $InGaO_3(ZnO)_m$ ($m \equiv 1$ to n) from 1985 to 2000 (1-5). In particular, $InGaZnO_4$ and $In_2Ga_2ZnO_7$ were synthesized in 1994. The pixels including field-effect transistors (FETs) formed using such an oxide semiconductor material were reported in 1992. Nomura et al. reported in 2003 that the oxide of IGZO had FET characteristics when it had a carrier concentration of lower than 10^{18}cm^{-3} (6, 7). At that time, the on/off ratio in the I_d-V_g characteristics was 6 to 7 orders of magnitude, which was low. Kawasaki and Ohno and Sugihara et al. increased the on /off ratio to 8 to 9 orders of magnitude by adding hydrogen to the oxide semiconductor (8, 9). We propose the application of oxide semiconductors not only to pixels of displays but also to a scan driver and a source driver in peripheral circuits in organic light-emitting diode (OLED) displays or liquid crystal displays (LCD); the source driver and the scan driver are integrated (10-28). Thus, it is possible to reduce the number of components used in the displays. Accordingly, reliability can be improved and cost can be reduced.

This time, we focus on the off-state current (I_{off}) of oxide semiconductor FETs, which is lower than that of amorphous silicon (a-Si) FETs. Power consumption is markedly reduced also in an LCD panel by extremely decreasing frame frequency, i.e., stopping unnecessary rewrites at the time of displaying still images. Thus, our LCD panel can be applied to an electronic paper display.

More cellular phones use extremely low-power reflective LCDs as sub displays. Our technique can also be applied to such LCDs. Moreover, the application of LCD panels solves the problems of electrophoretic displays and expands the potential of electronic paper displays.

In this paper, first, the characteristics of an IGZO FET are shown. Next, the specifications of our 6.0-in. extended graphics array (XGA) reflective LCD (RLCD) panel are shown. Then, we discuss a driving method for reducing power consumption. Finally, we discuss the display results and visual characteristics of our 6.0-in. XGA RLCD panel.

Characteristics of IGZO FET

Figure 1 shows the characteristics of our IGZO FET. The FET used for the measurement has a channel width W of 50 μm, a channel length L of 6 μm, and a threshold voltage V_{th} of 1.1 V. The FET is normally off; thus, we can integrate drivers with the panel. As disclosed in Reference 26, the I_{off} of the FET is 10^{-18} A/μm or lower. We have proposed the applications utilizing these characteristics. Also in this paper, we report the LCD panel utilizing these properties.

There is a concern that the reliability of oxide semiconductors at the time of light irradiation may decrease. We conducted the reliability test under the condition of an illuminance of 10 klx and a bias of ±20 V at 80 °C for 1 hour. As a result, our FET has comparatively high reliability in both +BT and −BT conditions. We therefore judged that our FET has high endurance against illumination and negative bias stress.

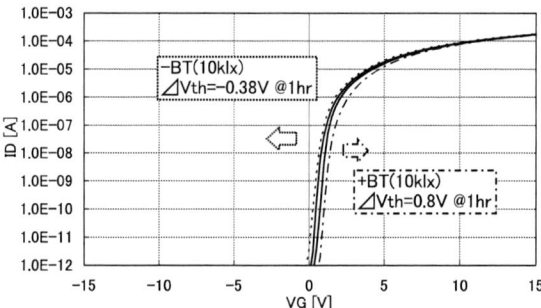

Figure 1. Characteristics of our IGZO FET in ±20 V BT condition (V_d = 5 V, W/L = 50 μm/6 μm, V_{th} = 1.1 V, illuminance = 10 klx).

Panel configuration

Table I shows the specifications of our 6.0-in. XGA RLCD panel using IGZO FETs.

Making use of the mobility of the oxide semiconductor, which is higher than that of a-Si, not only a scan driver but also a data selection demultiplexer is formed on a glass substrate (23-28).

In our panel, we use nematic liquid crystals generally used in active-matrix displays. The operation mode is a twisted nematic (TN) mode.

TABLE I. Panel Specifications

	Specifications
Screen Diagonal (in.)	6.0
Driving Method	Active Matrix
Resolution (pixels)	768 × 1024 (XGA)
Pixel Pitch (mm)	0.12 × 0.12
Pixel Density (ppi)	221
Aperture Ratio (%)	88.4
Pixel Arrangement	B/W square pixel
Source Driver	Demultiplexer
Scan Driver	Integrated

Driving method

From the point of view of application to electronic paper displays, lower power consumption is essential. The RLCD panel we fabricated this time operates at 60 fps at the time of displaying moving images like a normal LCD panel. However, by using the ultra-low I_{off} of the IGZO FET, we can markedly reduce the frequency of data rewrite when still images are displayed. We can therefore reduce the power consumption. This method can realize the electronic paper function.

First, we describe the specific operation of the driver. Figure 2 is a timing chart of power supply voltage and signals at the time of displaying still images. Not only the supply of signals but also the supply of high power supply voltage V_{DD} is stopped. In the driver, at the time of data rewrite, V_{DD} is supplied, a normal clock is supplied after clock lines become high concurrently, and then, a start pulse is input. After that, the data is completely rewritten. Then, the clock lines become low and the supply of V_{DD} is stopped. With such driving, power consumption will be reduced. For example, by changing a frequency of 60 fps into a frequency of 1/180 fps (data is rewritten once every 3 min.), the power consumption at the panel portion will be approximately 1/10000. If the power consumption can be reduced to this extent, our LCD panel will be applicable not only to the sub displays of cellular phones but also to electronic paper displays.

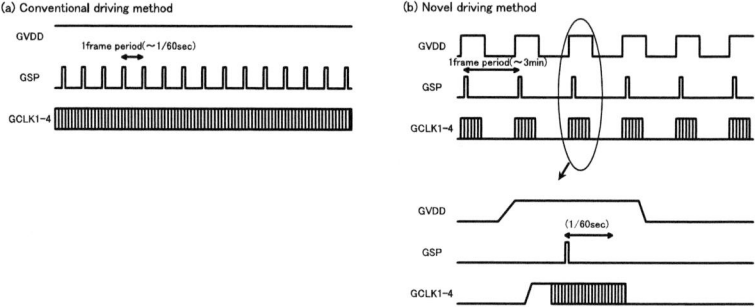

Figure 2. Timing chart of (a) conventional driving method and (b) novel driving method.

Results

We examined the extent to which frequency can be lowered in our RLCD panel. Figure 3 shows the luminance change in our novel driving method (at 1/600 fps). When a change in luminance of the panel is measured, the luminance hardly changed in the first 180 s after data was rewritten. Then, the luminance rose gradually until refresh operation. The results show that the interval between data rewrites can be extended to approximately 3 min. In the case where black and white still images such as books and documents are displayed, the interval between data rewrites can be further extended in terms of visibility.

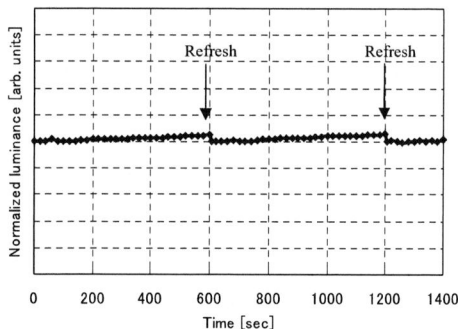

Figure 3. Luminance change in our novel driving method (at 1/600 fps).

Figure 4 shows the display picture of our 6.0-in. XGA LCD panel. Even when data was rewritten once every 3 min., the LCD panel was able to display images without problems.

Figure 4. Display picture.

Figure 5 shows power consumption at the time of displaying images at 60 fps and 1/180 fps. The power consumption at the time of displaying images at 1/180 fps was 1/10000 of that the power consumption at the time of displaying images at 60 fps.

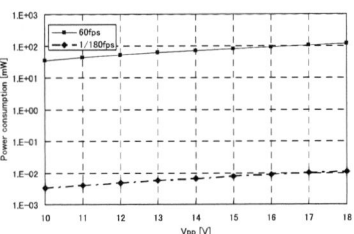

Figure 5. Power consumption of scan driver.

We evaluated the visual characteristics of the panel. Figure 6(b) shows the dependence of reflectance on the incident angle of light (θ) in the panel. Here, reflectance (R) is the average according to the azimuth angle (ϕ). Average reflectance is the relative reflectance obtained under the assumption that reflectance is 100% when a standard diffusing board is used. Electrophoretic displays have low incident-angle dependence. In our panel, the average reflectance has high-incident-angle dependence; however, when the incident angle is 35° or less, the reflectance of the panel is equal to or higher than that of the electrophoretic display. These characteristics of the panel are adequate for personal use.

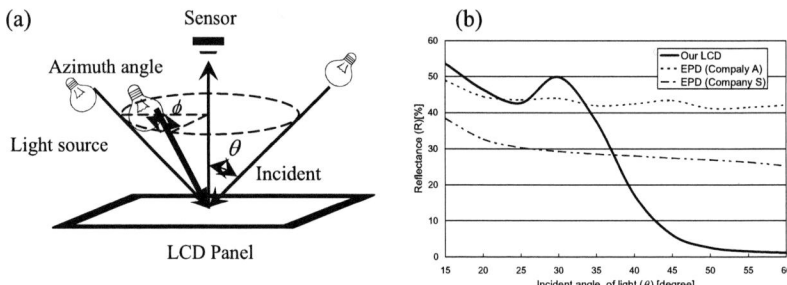

Figure 6. (a) Illustration of experiment and (b) Dependence of reflectance on incident angle of light.

Accordingly, our approach can solve the problems of electronic paper displays as follows: the drive voltage of drivers can be lowered; the difficulty of colorization and expression of gray levels is approximately the same as that of normal reflective LCDs; the response speed is sufficient; moving images can be easily displayed; and there is no problem with visibility. Our panel is very suitable for electronic paper displays.

Conclusions

We focused on the I_{off} of IGZO FETs, which is lower than that of a-Si FETs. We intended to reduce the power consumption of a LCD panel by significantly extending the interval between rewrites of still images, taking advantage of the extremely low I_{off} of IGZO FETs. In concrete terms, the interval between data rewrites could be extended to 3 min (1/180 fps). We found that normal LCDs could be applied to electronic paper displays. We actually fabricated a 6.0-in. XGA RLCD panel and determined its effectiveness.

References

1. N. Kimizuka and T. Mohri, Japanese Patents 1606567 (1991), 1639406 (1992), 1639407 (1992), 1606568 (1991), 1639398 (1992), and 1606571 (1991).
2. N. Kimizuka and T. Mohri, *J. Solid State Chem.* **60**, 382 (1985).
3. M. Nakamura, N. kimizuka and.T. Mohri, *J. Solid State Chem.* **93**, 298 (1991).
4. N. Kimizuka, M. Isobe and M. Nakamura, *J. Solid State Chem.* **116**, 170 (1995).
5. M. Nakamura, *NIRIM Newsl.* **150**, 1 (1995).
6. K. Nomura, H. Ohta, A. Takagi, T. Kamiya, M. Hirano, and H. Hosono, *Nature* **432**, 488 (2004).
7. K. Nomura, H. Ohta, K. Ueda, T. Kamiya, M. Hirano, and H. Hosono, *Science* **300**, 1269 (2003).
8. M. Kawasaki and H. Ohno, US Patent 6563174 (2003).
9. T. Sugihara, H. Ohno, and M. Kawasaki, WO2004/114391 (2004).
10. S. Maekawa and H. Kuwabara, Korean Patent 1022352 (2011).
11. K. Akimoto, T. Honda, and N. Sone, Chinese Patents 654582 (2010) and 668939 (2010).

12. K. Akimoto, T. Honda and N. Sone, US Patents 7674650 (2010), 7732819 (2010), 7910490 (2011), and 7932521 (2011).

13. K. Akimoto, Chinese Patent 731689 (2011).

14. H. Suzawa, S. Sasagawa, and T. Muraoka, US Patent 7915075 (2011).

15. T. Osada, K. Akimoto, T. Sato, M. Ikeda, M. Tsubuku, J. Sakata, J. Koyama, T. Serikawa, and S. Yamazaki, *SID Int. Symp. Dig. Tech. Pap.* **40**, 184 (2009).

16. H. Godo, D. Kawae, S. Yoshitomi, T. Sasaki, S. Ito, H. Ohara, A. Miyanaga, and S. Yamazaki, *SID Int. Symp. Dig. Tech. Pap.* **40**, 1110 (2009).

17. H. Ohara, T. Sasaki, K. Noda, S. Ito, M. Sasaki, Y. Toyosumi, Y. Endo, S. Yoshitomi, J. Sakata, T. Serikawa, and S. Yamazaki, *SID Int. Symp. Dig. Tech. Pap.* **40**, 284 (2009).

18. M. Takahashi, H. Kishida, A. Miyanaga, and S. Yamazaki, *IDW'08 Dig.*, 1640 (2008).

19. T. Osada, K. Akimoto, T. Sato, M. Ikeda, M. Tsubuku, J. Sakata, J. Koyama, T. Serikawa, and S. Yamazaki, *Jpn. J. Appl. Phys.* **43**, 03CC02 (2010).

20. H. Godo, D. Kawae, S. Yoshitomi, T. Sasaki, S. Ito, H. Ohara, H. Kishida, M. Takahashi, A. Miyanaga, and S. Yamazaki, *Jpn. J. Appl. Phys.* **43**, 03CB04 (2010).

21. H. Ohara, T. Sasaki, K. Noda, S. Ito, M. Sasaki, Y. Endo, S. Yoshitomi, J. Sakata, T. Serikawa, and S. Yamazaki, *Jpn. J. Appl. Phys.* **43**, 03CD02 (2010).

22. J. Sakata, H. Ohara, M. Sasaki, T. Osada, H. Miyake, H. Shishido, J. Koyama, Y. Oikawa, H. Maruyama, M. Sakakura, T. Serikawa, and S. Yamazaki, *Proc. IDW'09*, 689 (2009).

23. S. Amano, H. Harada, K. Akimoto, J. Sakata, T. Nishi, K. Moriya, K. Wakimoto, J. Koyama, S. Yamazaki, Y. Oikawa, T. Ikeyama, and M. Sakakura, *SID Int. Symp. Dig. Tech. Pap.* **41**, 626 (2010).

24. H. Shishido, K. Toyotaka, M. Tsubuku, K. Noda, H. Ohara, T. Nishi, K. Moriya, H. Godo, J. Koyama, S. Yamazaki, Y. Oikawa, T. Handa, and M. Sakakura, *SID Int. Symp. Dig. Tech. Pap.* **41**, 1128 (2010).

25. T. Nishi, K. Moriya, S. Fukai, Y. Kubota, K. Akimoto, J. Sakata, K. Kusunoki, R. Arasawa, K. Wakimoto, J. Koyama, S. Yamazaki, Y. Oikawa, K. Okazaki, and M. Sakakura, *SID Int. Symp. Dig. Tech. Pap.* **41**, 1685 (2010).

26. H. Godo, A. Miyanaga, K. Kusunoki, K. Toyotaka, T. Nishi, K. Moriya, S. Fukai, Y. Kubota, K. Wakimoto, J. Koyama, S. Yamazaki, Y. Oikawa, K. Okazaki, M. Sakakura, *Proc. IDW'10*, 235 (2010).

27. K. Toyotaka, K. Kusunoki, T. Nagata, Y. Hirakata, K. Wakimoto, J. Koyama, S. Yamazaki, R. Sato, K. Okazaki, and M. Sakakura, *Jpn. J. Appl. Phys.* **50**, 03CC09 (2011).

28. H. Shishido, S. Amano, K. Toyotaka, H. Miyake, T. Murakawa, T. Nishi, Y. Hirakata, J. Koyama, S. Yamazaki, K. Okazaki, T. Handa, and M. Sakakura, *SID Int. Symp. Dig. Tech. Pap.* (to be published).

96

Low Power 6.0-inch Extended Graphics Array Transmissive Liquid Crystal Display using Indium Gallium Zinc Oxide Semiconductor with Variable Frame Frequency

Seiko Amano[a], Hiroyuki Miyake[a], Takeshi Nishi[a], Yoshiharu Hirakata[a], Jun Koyama[a], Kenichi Okazaki[b], Masayuki Sakakura[b], and Shunpei Yamazaki[a]

[a] Semiconductor Energy Laboratory Co., Ltd., 398 Hase, Atsugi-shi, Kanagawa, 243-0036, Japan
[b] Advanced Film Device Inc., 161-2 Masuzuka, Tsuga-machi, Tochigi-shi, Tochigi, 328-0114, Japan

An oxide semiconductor thin film transistor (TFT) has much higher mobility than an amorphous silicon (a-Si) TFT, which leads to realization of larger displays with higher definition. Moreover, it has much lower off leak current than the a-Si TFT. We have focused on this feature and successfully fabricated a low-power 6.0-inch extended graphics array (XGA) transmissive liquid crystal display (LCD) with source and scan drivers integrated on a glass substrate. In order to realize low power consumption, unnecessary rewriting is not performed at the time of still image display. The interval between rewrite operations in our panel could be extended to about 1 min (= 1/60 fps) at the time of displaying still images. The power consumption of the drivers at 1/60 fps can be about 1/3,600 of that at 60 fps. We also discuss burn-in and flickers, which might be observed when this driving method is applied.

Introduction

Oxide semiconductors and displays including oxide semiconductors have been actively developed since the mid 1990's (1-8). We have investigated oxide semiconductors in recent years (9-24).

This time, we focus on the off leak current (I_{off}) of oxide semiconductor FETs, which is lower than that of amorphous silicon (a-Si) FETs. In order to realize low power consumption, the frame frequency is extremely decreased, that is, unnecessary rewriting is not performed when still images are displayed. With an external driver circuit judging whether a display image is a still image or a moving image, this driving is compatible with normal display of moving images. This method can be applied to devices that often display still images, such as mobile phones, digital photo frames, and PC monitors. With lower frame frequency, screen burn-in might occur when the same image is displayed for a long time. We therefore examined screen burn-in this time. We also discuss the flickers at the time of data rewriting.

First, we show characteristics of indium gallium zinc oxide (IGZO) FETs and the specifications of our 6.0-inch XGA transmissive LCD panel. Then, we explain a driving method for making the frame frequency variable. Finally, we provide the display results of our 6.0-inch XGA transmissive LCD panel and discuss conceivable problems like burn-in and flickers.

97

Characteristics of IGZO FETs

Figure 1 shows the characteristics of the IGZO FET. The FET has a channel width (W) of 50 μm, a channel length (L) of 6 μm, and a threshold voltage of 1.1 V. The FET is normally off; thus, we can integrate the drivers with the panel. As described in Ref. (24), I_{off} of the FET is 10^{-18} [A/μm] or lower.

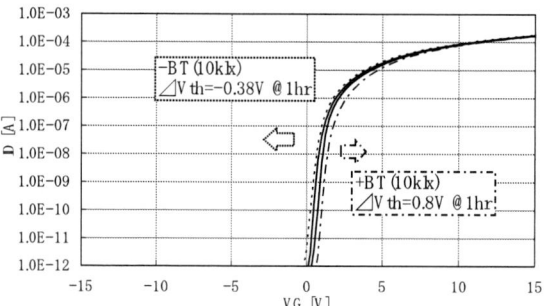

Figure 1. Characteristics of IGZO FET before and after BT (bias temperature) stress test. The test is performed under the conditions of +20/-20, 80 °C, and 10 klx (V_d =5 V, W/L = 50 μm/6 μm, V_{th} = 1.1 V). Dashed-dotted line and dotted line represent the results of +BT test and -BT test, respectively.

Figure 1 also shows the reliability of our FET under visible light irradiation. The results of the reliability test are obtained by the bias temperature (BT) test conducted by applying a bias of ±20 V to a gate electrode at 80 °C for 1 h with 10 klx light. The results prove that our FET has comparatively high reliability in both +BT and -BT conditions.

Panel Configuration

Table I shows the specifications of the 6.0-inch XGA transmissive LCD panel we fabricated using the IGZO FETs. The resolution of the panel is 212 ppi. Utilizing the mobility of the IGZO FETs, which is much higher than that of a-Si FETs, we formed not only a scan driver but also a source driver on a glass substrate. Figure 2 shows the circuit diagram and the timing chart of the scan driver. The source driver has a similar configuration. Since the IGZO FET we have developed is a normally-off type, we can use such drivers shown in Figure 2.

TABLE I. Panel Specifications

Specifications	
Screen Diagonal	6.0-inch
Resolution	768 x RGB x 1024 (XGA)
Pixel Pitch, Pixel Density	0.12mm (RGB) x 0.12mm, 212ppi
Aperture Ratio	54%
Source/Gate Drivers	both integrated

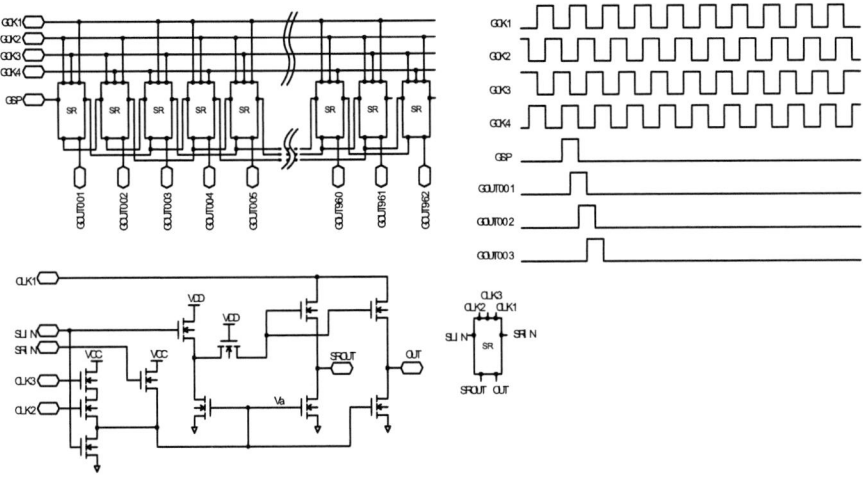

Figure 2. Circuit diagram and timing chart of scan driver

Driving Method

We examined how much frequency can be lowered in the transmissive LCD panel we formed this time.

First, Figure 3 shows timing charts of power supply voltage and signals at the time of displaying still images. Not only the supply of signals but also the supply of high power supply voltage (V_{DD}) is stopped in our novel method. In order to rewrite data on the screen with the driver employed here, V_{DD} is supplied first, and then, a normal clock is supplied after clock lines are simultaneously set high, and a start pulse is input. Then, data are completely rewritten, and then, the clock lines are set low and the supply of V_{DD} is stopped.

Such driving is expected to reduce power consumption. Table II shows the relation between the number of rewrite operations and calculated values of power consumption. For example, when data are rewritten once every 1 min (i.e., at 1/60 fps), the power consumption at the panel portion will be reduced to about 1/3,600 of the power consumption at 60 fps.

In this driving, normal driving of moving images can be combined with display of still images with a smaller number of rewrite operations. As described in Ref. (19), it is possible to perform display in accordance with input data of a moving image and input data of a still image.

In addition, when modifying the scan driver to provide a function of partial scan, we can apply this driving method even in the case that display of only part of the screen is changed (20). Thus it is expected that power consumption can be further reduced.

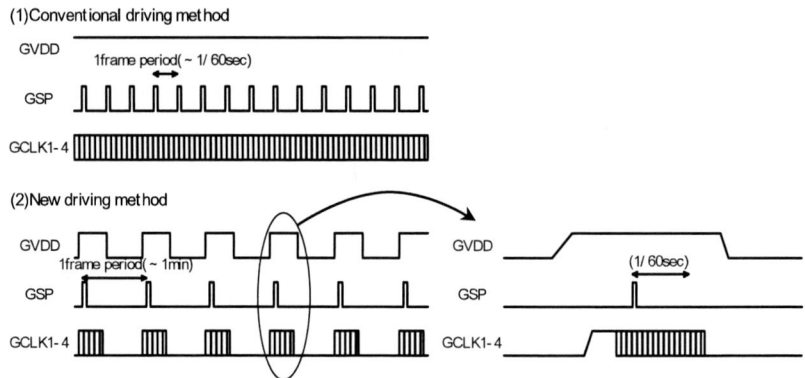

Figure 3. Timing chart of our driving method

TABLE II. The dependency of the power consumption of the scan driver on the frame frequency

Frame Frequency [fps]	120	60	1	1/60
Power Consumption [mW]	120	60	1	0.02

Results

Figure 4 is a display picture of the prototype of the 6.0-inch XGA transmissive LCD panel. Figure 5 shows the luminance change over time in our new driving method. The results show that the interval between rewrite operations can be extended to at least about 1 min. In the case of displaying monochrome still images such as documents, images are displayed with two gray levels having a large luminance difference, so that the interval between rewrite operations can be further extended in terms of visibility.

The measured value of the actual power consumption of the driver at that time was 30 µW, which is approximately 1/3,600 of the power consumption at the time of normal display at 60 fps.

We examined screen burn-in on a LCD. A checkered pattern shown in Figure 6(a) was displayed first at a frame frequency of 1/60 Hz for 10 min, and then a gray image was displayed. Burn-in was not observed as shown in Figure 6(b). Alternate-current driving with lower frame frequency is not likely to cause burn-in after the same image is displayed for a long time.

Finally, we discuss flickers, which are caused by noise of pulses that are sequentially output from the scan driver when data are rewritten. The voltage magnitude of the scan driver in our LCD using IGZO FETs can be much smaller than that of an a-Si TFT LCD because the mobility of the IGZO FET is much higher than that of the a-Si TFT. The magnitude of the noise is therefore also small. Owing to this, flickers are less observable in our LCD.

Figure 4. 6.0-inch display picture

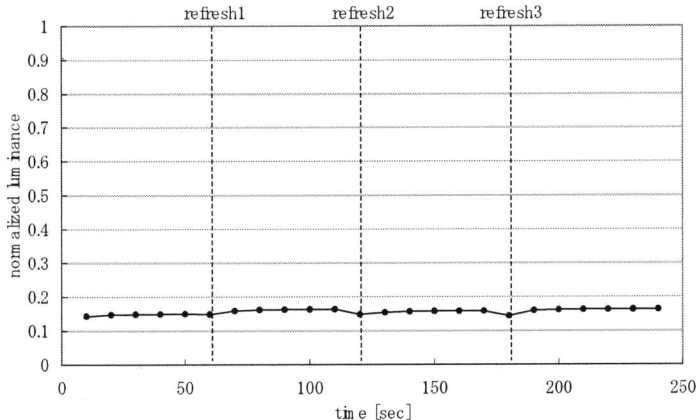

Figure 5. Luminance change in our new driving method

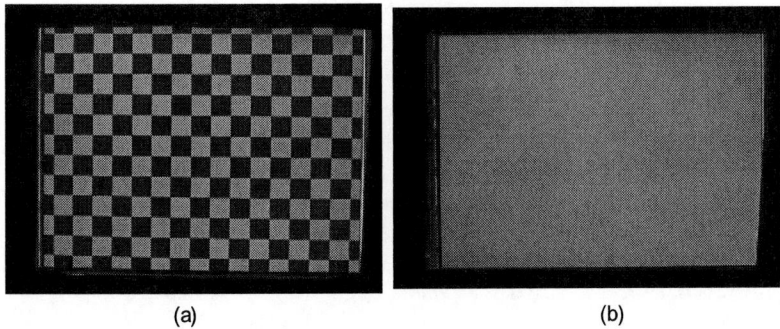

Figure 6. (a) Display of checkered pattern and (b) Gray screen after checkered pattern is displayed for 10 min

Conclusions

The interval between rewrite operations in the panel we fabricated this time could be extended to 1 min (= 1/60 fps) at the time of displaying still images, so that power consumption of the driver could be drastically reduced. We also succeeded in combining this novel driving with normal driving of moving images. Although we demonstrated the 6.0-inch XGA transmissive LCD panel, we can also apply this driving method to PC monitors of over 20 inches. Moreover, we showed that alternate-current driving at 1/60 Hz does not cause burn-in even when the same image continues to be displayed for about 10 min.

References

1. N. Kimizuka and T. Mohri, *J. Solid State Chem.*, **60**, 382 (1985)
2. N. Kimizuka, M. Isobe, and M. Nakamura, *J. Solid State Chem.*, **116**, 170 (1995)
3. N. Kimizuka and T. Mohri, Japanese Patents 1606567 (1991), 1606568 (1991), 1606571 (1991), 1639398 (1992), 1639406 (1992), and 1639407 (1992)
4. M. Nakamura, *NIRIM Newsletter*, **150**, 1 (1995)
5. M. Nakamura, N. Kimizuka, and T. Mohri, *J. Solid State Chem.*, **93**, 298 (1991)
6. C. Li, Y. Bando, M. Nakamura, M. Onoda, and N. Kimizuka, *J. Solid State Chem.*, **139**, 347 (1998)
7. K. Nomura, H. Ohta, A. Takagi, T. Kamiya, M. Hirano, and H. Hosono, *Nature*, **432**, 488 (2004)
8. K. Nomura, H. Ohta, K. Ueda, T. Kamiya, M. Hirano, and H. Hosono, *SCIENCE*, **300**, 1269 (2003)
9. S. Maekawa and H. Kuwabara, Korean Patent 1022352 (2011).
10. K. Akimoto, T. Honda, and N. Sone, Chinese Patents 654582 (2010) and 668939 (2010).
11. K. Akimoto, T. Honda and N. Sone, US Patents 7674650 (2010), 7732819 (2010), 7910490 (2011), and 7932521 (2011).
12. K. Akimoto, Chinese Patent 731689 (2011).
13. H. Suzawa, S. Sasagawa, and T. Muraoka, US Patent 7915075 (2011).
14. T. Osada, K. Akimoto, T. Sato, M. Ikeda, M. Tsubuku, J. Sakata, J. Koyama, T. Serikawa, and S. Yamazaki, *Jpn. J. Appl. Phys.*, **49**, 03CC02-1 (2010)
15. H. Godo, D. Kawae, S. Yoshitomi, T. Sasaki, S. Ito, H. Ohara, H. Kishida, M. Takahashi, A. Miyanaga, and S. Yamazaki, *Jpn. J. Appl. Phys.*, **43**, 03CB04-1 (2010)
16. H. Ohara, T. Sasaki, K. Noda, S. Ito, M. Sasaki, Y. Endo, S. Yoshitomi, J. Sakata, T. Serikawa, and S. Yamazaki, *Jpn. J. Appl. Phys.*, **43**, 03CD02-1 (2010)
17. M. Takahashi, H. Kishida, A. Miyanaga, and S. Yamazaki, *IDW'08 Digest*, 1637 (2008)

18. J. Sakata, H. Ohara, M. Sasaki, T. Osada, H. Miyake, H. Shishido, J. Koyama, Y. Oikawa, H. Maruyama, M. Sakakura, T. Serikawa and S. Yamazaki, *IDW'09 Proc.*, 689 (2009)

19. S. Amano, H. Harada, K. Akimoto, J. Sakata, T. Nishi, K. Moriya, K. Wakimoto, J. Koyama, S. Yamazaki, Y. Oikawa, T. Ikeyama, and M. Sakakura, *SID Symposium Digest*, **41**, 626 (2010)

20. H. Harada, A. Umezaki, S. Amano, T. Nishi, K. Moriya, K. Wakimoto, J. Koyama, S. Yamazaki, Y. Oikawa, T. Ikeyama, and M. Sakakura, *Proc. AM-FPD'10 Digest*, 199 (2010)

21. H. Shishido, K. Toyotaka, M. Tsubuku, K. Noda, H. Ohara, T. Nishi, K. Moriya, H. Godo, J. Koyama, S. Yamazaki, Y. Oikawa, T. Handa, and M. Sakakura, *SID Symposium Digest*, **41**, 1128 (2010)

22. T. Nishi, K. Moriya, S. Fukai, Y. Kubota, K. Akimoto, J. Sakata, K. Kusunoki, R. Arasawa, K. Wakimoto, J. Koyama, S. Yamazaki, Y. Oikawa, K. Okazaki, and M. Sakakura, *SID Symposium Digest*, **41**, 1685 (2010)

23. K. Toyotaka, K. Kusunoki, T. Nishi, K. Moriya, S. Fukai, Y. Kubota1, K. Wakimoto, J. Koyama, S. Yamazaki, Y. Oikawa, K. Okazaki, and M. Sakakura, *Proc. AM-FPD'10 Digest*, 17 (2010)

24. H. Godo, A. Miyanaga, K. Kusunoki, K. Toyotaka, T. Nishi, K. Moriya, S. Fukai, Y. Kubota, K. Wakimoto, J. Koyama, S. Yamazaki, Y. Oikawa, K. Okazaki, and M. Sakakura, *IDW'10 Proc.*, 235 (2010)

104

Comparative Analysis of Organic Thin Film Transistor Structures for Flexible E-Paper and AMOLED Displays

Linrun FENG, Xiaoli XU and Xiaojun GUO

Department of Electronic Engineering, Shanghai Jiao Tong University, Shanghai, 200240, China

Organic thin-film transistors (OTFTs) have attracted considerable attention in applications for driving flexible e-paper and active matrix organic light-emitting diode (AMOLED) displays. In the systems, the pixel electrode, which connects the bottom electrode of the display media to the drain or source of the switch OTFT in the e-paper display, and the driving OTFT in the AMOLED display, will form parasitic effects with the OTFTs. Through numerical device simulations, it is found that, in the bottom-contact bottom-gate (BCBG) structure OTFT backplane, the presence of the pixel electrode may result in a shift of the transfer characteristics and a significant decrease of the output impedance. Although the DC electrical characteristics of the bottom-contact top-gate (BCTG) OTFTs are not affected by the presence of the pixel electrode, the BCTG structure has a larger parasitic capacitance, which can cause a higher feed-through voltage for performing switching in the e-paper displays.

Introduction

Owning to their excellent intrinsic flexibility and capability of being manufactured by cost-effective solution or printing processes at a low temperature, organic thin-film transistors (OTFTs) have attracted considerable attention in applications for driving flexible e-paper and active matrix organic light-emitting diode (AMOLED) displays. In the last decade, great efforts have been paid to develop high carrier mobility, chemically and physically stable organic semiconductor materials, which can now meet general requirements of driving e-paper and AMOLED displays. It has also been well proved that the device structure can significantly affect the OTFT's electrical performance (1, 2). Generally, the top-contact (TC) structure OTFTs can form better contacts between the metal electrode and the semiconductor layer than the bottom-contact (BC) ones to provide more efficient charge injection (3). However, bottom-contact structures are still more popularly used in circuit integration of OTFTs for display backplanes because of the process difficulty of making source/drain metal contacts onto the organic semiconductor layer with precise patterning (4-6). With respect of the relative locations of the gate electrode and the source/drain electrodes, BC OTFTs can be realized in two configurations: BC top gate (BCTG, also named inverted staggered structure), and BC bottom gate (BCBG, also named coplanar structure). Both structures have been widely employed in applications of display backplanes depending on process integration preferences (7, 8).

As shown in Fig. 1, in e-paper display applications, the OTFT performs as a switch, while in AMOLED displays, two types of OTFTs are required for switching and current driving functions, respectively. The switching OTFTs in both displays need to have a high enough on-off current ratio to charge the pixel to the operational voltage within the line selection period and to hold the charge on the pixel until the display is refreshed. The driving OTFTs in AMOLED displays are needed to provide uniform and stable current, as analog current sinks or current sources, to drive the OLEDs (9, 10).

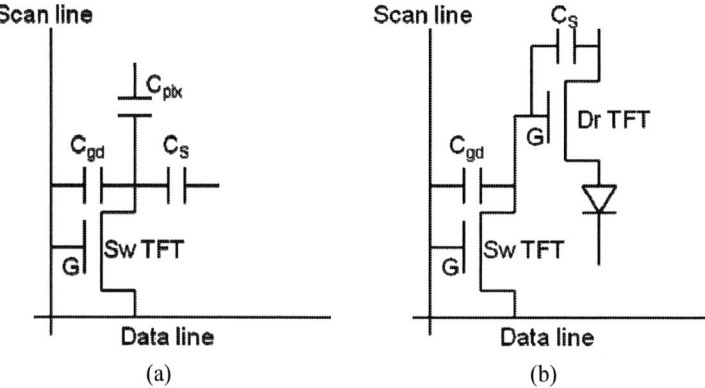

(a)   (b)

Figure 1. Typical pixel circuits for (a) the e-paper display and (b) the AMOLED display.

To integrate the OTFT backplane with the front plane of display media for a completed display system, the pixel electrode, which works as the bottom electrode of the front plane, is connected to the drain or source of the switching OTFT in the e-paper display, and the driving OTFT in the AMOLED display. Since the pixel electrode is directly on top of the OTFTs separated by an interlayer dielectric (ILD) layer of a certain thickness, the resulted parasitic capacitance between the pixel electrode and the intrinsic part of the OTFTs may affect their functions of switching or current driving. In this work, the parasitic effects caused by pixel electrode in both BCBG and BCTG OTFT backplanes will be carefully investigated and compared.

Simulation Methods

Two-dimensional numerical simulations were performed using the commercial software Atlas vended by SILVACO in this study to exclude any influence of process induced effects (11). Although originally developed for silicon and inorganic devices, Atlas allows user-defined semiconductor materials and has been proved to be a useful tool for studying device physics of OTFTs (1).

The BCBG and BCTG device structures used in the simulations are given in Fig. 2, with the pixel electrode being electrically connected to the drain electrode. The devices have a 50 nm thick channel with the length of 10 μm, 100 nm thick source/drain electrodes with the length of 10 μm, and a gate insulator of 300 nm with the dielectric constant of 4.0. The thickness of the interlayer dielectric (t_{ILD}) between the pixel electrode and the intrinsic part of the OTFTs is varied, and the dielectric constant is also set to be 4.0. The channel width is 1 μm.

The usual values of pentacene (energy gap of 2.5 eV, ionization potential of 5eV and dielectric constant of 4.0) have been used for the organic semiconductors as the channel

(12). The field-dependent hole mobility is described by the Poole-Frenkel model, which can be expressed as:

$$\mu = \mu_0 e^{\sqrt{E/E_0}}$$ [1]

where μ_0, the low field mobility, is set to be 0.033 cm^2/V·S, E is the electrical field and E_0 is a characteristics parameter equal to 3×10^5 V/cm. The effective density of the states (N_V) is set to be 10^{-19} cm^{-3}. Simulations based on these models and parameters have been proved to be able to get the results well fitting with the experimental data (12).

Neither bulk semiconductor trap states nor interfacial trap states have been included in the simulations, which will not affect the qualitatively comparative study in this work.

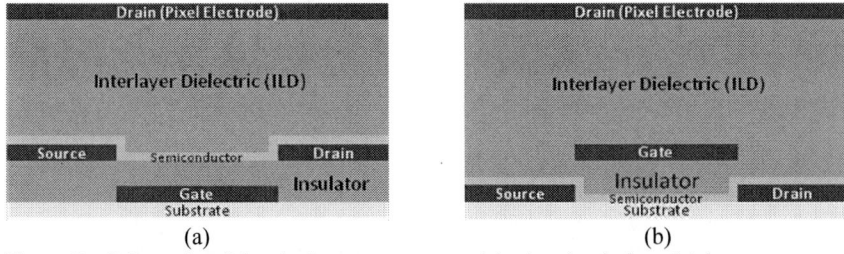

(a)   (b)

Figure 2. Schematic of the device structures used in the simulation: (*a*) bottom-contact bottom-gate (BCBG) structure with the pixel electrode and (*b*) bottom-contact top-gate (BCTG) structure with the pixel electrode.

By varying t_{ILD}, the switching and current driving performance of BCBG and BCTG OTFTs are fully investigated and compared for applications in e-paper and AMOLED displays.

Results and Discussions

Switching Performance

The simulated transfer characteristics (I_{ds}-V_{gs}) of both BCBG and BCTG OTFTs at different t_{ILD} are given in Fig. 3. For the BCBG OTFT, it's obviously observed from Fig.3 (a) that the presence of the pixel electrode results in a shift of the transfer characteristics to the positive as the drain bias is increased. The magnitude of the shift decreases with the increase of the t_{ILD}. The transfer characteristics of the BCTG OTFT are not affected by the presence of the pixel electrode, as shown in Fig. 3 (b), which is attributed to the electrical shielding of the channel from the effects of the pixel electrode by the top gate.

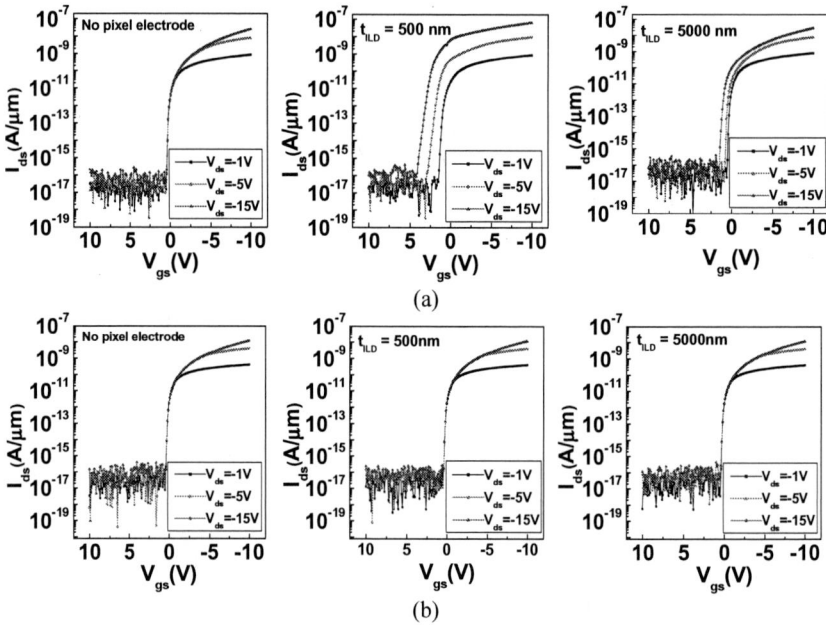

Figure 3. Simulated transfer characteristics of the OTFTs in (*a*) BCBG structure and (*b*) BCTG structure, without the pixel electrode, and with the pixel electrode in the cases of interlayer dielectric thickness (t_{ILD}) of 500 nm and 5000 nm respectively.

In the BCBG structures, the pixel electrode covers the channel region and acts as the second gate, which makes the device similar to a double-gate transistor. As a drain bias is added, an additional conductive channel will be formed near the interface between the ILD layer and the organic semiconductor layer, as shown in Fig. 4, and become more conductive with the increase of the drain bias. Therefore, there is a shift of the transfer characteristics and the drain current at the same gate bias voltage is increased. The phenomena have also been experimentally demonstrated in double-gate OTFTs, where the presence of the second gate with a certain voltage bias can cause a shift of the threshold voltage (13, 14). The increase of t_{ILD} will reduce the influence of the pixel electrode, and can thus bring a smaller shift of the I_{ds}-V_{gs} characteristics.

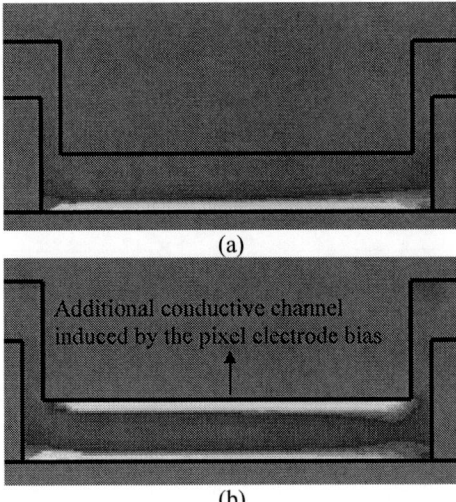

(a)

Additional conductive channel
induced by the pixel electrode bias

(b)

Figure 4. Schematic diagrams of the formed conductive channels in the BCBG structure OTFTs: (*a*) without the pixel electrode and (*b*) with the pixel electrode. Both the gate and drain biases are -8V

When the OTFT works as a switch in e-paper or AMOLED displays, a high enough on-off current ratio should be met. For the BCTG structure, the presence of the pixel electrode does not cause any changes to the I_{ds}-V_{gs} electrical characteristics, so no additional design consideration is needed. But for the BCBG structure, according to the results in Fig. 3, a wider gate voltage swing must be used to enable the device to be turned on and off as required. The exact value of the voltage swing depends on how much the shift of the I_{ds}-V_{gs} characteristics is, which is a function of both the maximum input data voltage and t_{ILD}.

For the application of the OTFT as a switch, a small feed-through voltage is also important. The feed-through voltage (ΔV_{ft}) can be expressed as:

$$\Delta V_{ft} = V_{gw} \times C_{gd}/(C_{gd} + C_s)$$ [2]

where V_{gw} is the gate voltage swing, C_{gd} is the gate-to-drain parasitic capacitance and C_S is the storage capacitance (as shown in Fig. 1).

Based on [2], a larger C_S can be designed to reduce ΔV_{ft}, which, however, is limited by the pixel area and the strict requirement of the fast charging time. Therefore, as seen from equation [2], it's vital to minimize the C_{gd} in the design to reduce ΔV_{ft}. The simulated characteristics of gate-to-drain parasitic capacitance C_{gd} as a function of V_{gd} are given in Fig. 5, the drain electrode is zero biased, for both BCBG and BCTG structures with different values of t_{ILD}.

In the whole operation regimes, the BCBG structure owns smaller C_{gd} than the BCTG one. For the BCTG structure, additional capacitance is formed between the pixel electrode and the gate electrode, therefore a larger C_{gd} is induced, and increases with the decrease of t_{ILD}. When the OTFTs are operated in the ON state with a negative gate bias, C_{gd} in the BCBG structure does not change with the presence of the pixel electrode and the variations of t_{ILD}, since the pixel electrode is shielded from the gate by the conductive channel. When the OTFTs are turned off by a positive gate bias, C_{gd} of both structures increases as the t_{ILD} decreases. For the BCBG structure, since there is no conductive channel formed in the OFF state, an additional capacitance is induced between the pixel and gate electrode through the multi-layer dielectric composed of the ILD layer, organic semiconductor layer and the gate insulator layer.

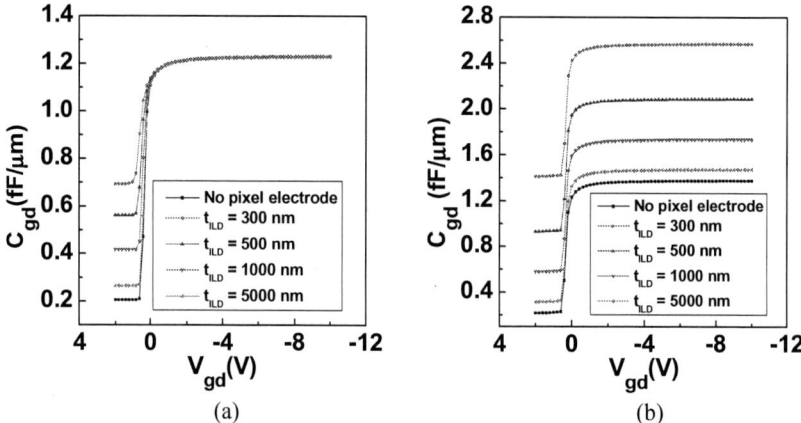

(a)   (b)

Figure 5. The simulated gate-to-drain parasitic capacitance (C_{gd}) as a function of V_{gd} for (a) BCBG structure and (b) BCTG structure OTFTs, without the pixel electrode, and with a pixel electrode in the cases of interlayer dielectric thickness (t_{ILD}) of 300, 500, 1000 and 5000 nm.

In a summary for this part, to act as a switch in e-paper or AMOLED displays, the BCTG OTFT structure owns the advantage of no shift of I_{ds}-V_{gs} characteristics with the presence of the pixel electrode, but induces a larger C_{gd}. To reduce the parasitic capacitance, a thicker ILD layer is required. For the BCBG OTFT structure, the C_{gd} is much smaller, but the design needs to increase the gate voltage swing considering the shift of the I_{ds}-V_{gs} characteristics with the increase of the drain bias.

Current Driving Performance

When the OTFT works as a current source or sink in the AMOLED display, a high output impedance in the saturation regime is needed to provide a stable current without being affected by variations of the drain bias. As stated in the above, for the BCTG structures, since the channel is shielded from the pixel electrode by the gate, the output impedance is not affected by the pixel electrode. The following will mainly discuss the case for the BCBG structure. Fig. 6 shows the effects of the pixel electrode on the output characteristics of the BCBG OTFT at different t_{ILD}. The output impedance is degraded with the presence of the pixel electrode, due to the additional channel formed at the interface between the ILD layer and the organic semiconductor layer, as already illustrated in Fig. 4. Even when the t_{ILD} is increased to 5000 nm, there is still a significant decrease of the output impedance compared to that without the pixel electrode.

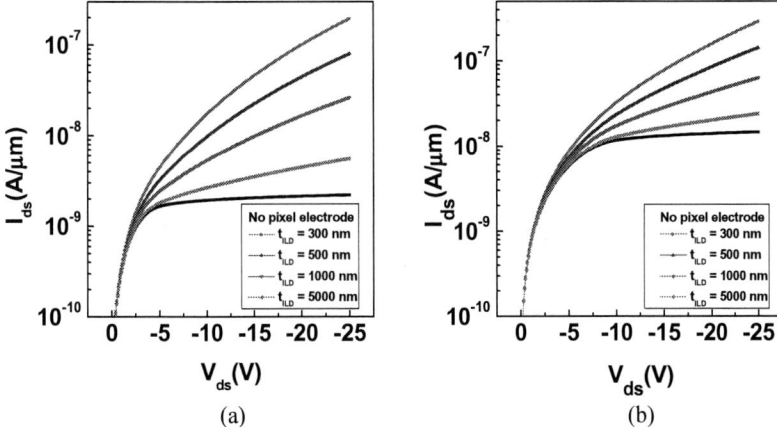

(a)   (b)

Figure 6. The simulated output characteristics of the BCBG structure OTFT without the pixel electrode, and with the pixel electrode in the cases of interlayer dielectric thickness (t_{ILD}) of 300, 500, 1000 and 5000 nm: (a) V_{gs}=-4V and (b) V_{gs}=-8V .

Therefore, for the current driving application in the AMOLED display, the BCBG structure suffers the degraded output impedance due to the presence of the pixel electrode, while the BCTG structure does not have this issue. To use the BCBG OTFT for high-performance AMOLED displays, a very thick ILD layer is required to effectively suppress the parasitic effects, which, however, may increase the process difficulty to form the via hole connecting the pixel electrode and the drain electrode of the OTFT.

Conclusions

In this work, a comparative analysis of BCBG and BCTG OTFT structures for e-paper and AMOLED displays has been carried out. In the BCBG structure backplane, the presence of the pixel electrode results in a shift of the transfer characteristics and degradation of the output impedance of the OTFTs. When the device works as a switch in the e-paper display, a wider gate voltage swing can be designed to compensate the shift of the transfer characteristics. But for AMOLED drive OTFT applications, a very thick ILD layer is needed to effectively suppress the parasitic effects induced by degradation of the output impedance, which, however, might increase the process difficulty. In the BCTG structure backplane, the electric characteristics of the OTFTs are not affected by the presence of the pixel electrode, but a larger parasitic capacitance C_{gd} can cause a higher feed-through voltage for performing switching in the e-paper displays, which needs to be considered in the design.

Acknowledgments

The work is supported by The Program for Professor of Special Appointment (Eastern Scholar) at Shanghai Institutions of Higher Learning, Program for New Century Excellent Talents (NCET) in University in China, and the NSFC of China (Grant No. 60906039, 60910295).

References

1. C. H. Shim, et al., *IEEE Trans. Electron Devices*, **57**, 195 (2010)
2. D. J. Gundlach, et al., *Journal of Applied Physics*, **100**, 024509 (2006)
3. I. G. Hill, *Appl. Phys. Lett.*, **87**, 163505 (2005)
4. J. Yuan, et al., *Appl. Phys. Lett.*, **82**, 3967 (2003)
5. M. Mizukami, et al., *IEEE Electron Device Letters* **27**, 249 (2006)
6. H. Yan, et al., *Appl. Phys. Lett.*, **87**, 183501 (2005)
7. S. E. Burns, et al., *Journal of the SID*, **13**(7), 583 (2005)
8. I. Yagi, et al., *Journal of the SID* **16**(1), 15 (2008)
9. R. A. Street, *Adv. Mater.*, **21**, 2007 (2009)
10. G. Gelinck, et al., *Adv. Mater.*, **22**, 1 (2010)
11. ATLAS User's Manual, Silvaco Int. Inc., Santa Clara, CA, (2005)
12. A. Bolognesi, et al., *IEEE Trans. Electron Devices* **51**, 1997, (2004)
13. G. H. Gelinck, et al., *Appl. Phys. Lett.*, **87**, 073508 (2005)
14. S. Iba, et al., *Appl. Phys. Lett.*, **87**, 023509 (2005)

CHAPTER 3

CHALLENGES IN GRAPHENE, NANOWIRE,
AND NANOTUBE DEVICES

114

Recent Progress in Facile Preparation of Graphene

Z. Y. Lin[a], Y. G. Yao[a], Z. Li[a], K. S. Moon[a], and C. P. Wong[a,b]

[a] School of Materials Science & Engineering, Georgia Institute of Technology, Atlanta, Georgia 30332, USA
[b] Faculty of Engineering, The Chinese University of Hong Kong, Hong

> The preparation of graphene is a major topic of interest in graphene research. We report a rapid reduction of graphene oxide (GO) by microwave radiation, and a mild solvothermal method for controllable reduction. These methods are useful for producing reduced GO for supercapacitor electrode materials. Moreover, we also demonstrate the single layer graphene growth in an atmosphere CVD system on a Cu substrate using a liquid carbon precursor.

Introduction

Graphene is a single layer of sp^2 carbon atoms covalently boned in a two-dimensional honeycomb lattice. Since its discovery at 2004 (1), lots of researches have been focused on graphene due to the outstanding material properties, including high charge carrier mobility (200,000 $cm^2V^{-1}s^{-1}$)(2), high thermal conductivity ($\sim$5,000 $W\ m^{-1}K^{-1}$) (3), optical transparency(2.3% absorption of white light) (4), mechanical strength (Young's modulus, $\sim$1100 GPa) (5), and high surface area ($\sim$2630 m^2/g, calculated value) (6). These properties lead to very promising applications of graphene in electronic devices, transparent electrodes, heat dissipation, reinforce materials in composites as well as energy storage devices etc.

The performance of graphene in these proposed applications strongly relies on the preparation method, which is a major topic of the recent interest. Many methods have been successfully used in producing graphene, including micromechanical exfoliation of highly oriented pyrolytic graphite (also called scotch tape method) (1), ultrasonication exfoliation of graphite (7), epitaxial growth from SiC substrate (8), chemical vapor deposition (CVD) on metal substrate (9), reduction of graphene oxide (GO) (10), etc. However, issues of high cost, low yield as well as defective graphene product limit the practical applications of these methods.

Here we will introduce recent progresses in our group toward facile preparation of graphene by reduction of GO as well as CVD growth. A microwave (MW) reduction of GO has been developed which is able to produce graphene from GO within seconds of MW radiation. Moreover, a low temperature thermal reduction method is systematically studied to demonstrate the reduction of GO with a controllable degree of reduction. The performance of reduced GO (rGO) from these methods as the supercapacitor electrode material has been evaluated. Furthermore, we also work on the CVD growth of graphene in a low-cost atmosphere CVD system.

Reduction of GO

The reduction of GO is the most promising method for mass production of graphene. GO is a non-stoichiometric material prepared by oxidation of graphite. The structure of GO is of considerable debate, but it is generally accepted that oxygen-containing functional groups are dispersed on the graphene sheet, including carboxyls, carbonyls, epoxides and hydroxyls (11). Successful reduction of GO will have the oxygen-containing functional groups removed and the conjugated system restored. The focus of our research is to developing a low-cost, large-scale as well as green method for GO reduction.

MW reduction of GO

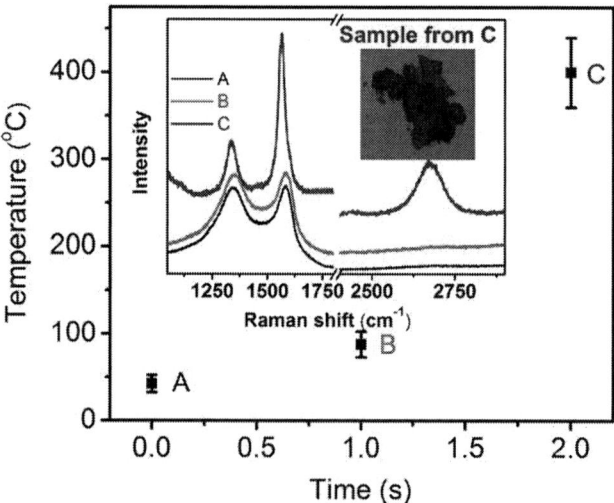

Figure 1. The temperature profile of GO film under 6.425 GHz, 500 W MW. Inset is Raman spectra of rGO as well as a picture of rGO from point C.

We developed a rapid, dry approach to synthesize rGO by direct MW radiation of GO film (12). Figure 1 shows the temperature profile during MW treatment. It can be seen that the temperature rises up very rapidly to ~ 400 °C within 2 seconds. The MW absorbing properties of GO may be due to the combination of the polar functional groups in GO, and the sp2 conjugated structure which is similar to graphite and carbon nanotubes. We believe that the thermal heating effect plays an important role in the reduction of GO. Dark powder was obtained after MW reduction, as shown in the inset of Figure 1. Moreover, it was found that the bulk conductivity increased from 0.07 S/m to 1×10^4 S/m after the MW treatment, close to thermally exfoliated graphene at 1050 °C. As shown in the inset of Figure 1, the D/G ratio from the Raman spectra decrease, indicating a decreased amount of defects in rGO. Moreover, the 2D peak at ~2700 cm^{-1} is present in the spectrum of rGO (Point C in Figure 1), which is very impressive and indicates the formation of graphene structure. The reduction of GO is further confirmed by XPS, TGA and other techniques.

The supercapacitive behavior of rGO from MW reduction was characterized in a three electrode system using 1 M H_2SO_4 as the electrolyte (13). It was found that the specific capacitance of MW rGO was ~147.5 F/g at a discharge current of at 10 mA/cm^2, which is ~1.6 times that of the commercial activated carbon (87.8 F/g).

Solvent-assisted thermal reduction of GO

Low temperature thermal reduction of GO is of particular interest for graphene applications that requires tunable degree of reduction as well as good dispersion. It has been found that GO dispersion can be reduced at a temperature as low as 100 °C. Figure 2 shows the images that the color of GO/water dispersion changes from light yellow to dark black after 24 hours' reflux. The temperature plays an important role here because that GO/dimethylformamide (DMF) dispersion at 150 °C shows a much quicker color change, as seen in Figure 2. This color change is usually considered as a sign of reduction, due to the restoration of conjugated system. The successful reduction of GO in these conditions is also confirmed by UV-visible, TGA, XPS, etc. Furthermore, from a set of structural characterization, it was found that at temperatures of 100 and 150 °C the carboxylic and carbonyl groups decomposed (14).

An interesting phenomenon is observed during the sovlothermal reduction that reduction rate is highly dependent on the solvent used (14). At 150 °C, DMF accelerates the GO reduction rate significantly, while dimethyl Sulphoxide (DMSO) has less acceleration effect. However, ethylene glycol (EG) reduces the reduction rate compared to dry conditions. The solvent-GO interaction, including polar-polar interaction and hydrogen bonding may be one possible reason for the solvent-dependent reduction.

The low temperature solvothermal method allows a fine control of the degree of reduction of GO. We have demonstrated that the C/O ratio can be continuously adjusted between ~ 2 - ~ 6 by simply changing the treatment time (15). We studied the supercapacitive performance of rGO with different degrees of reduction. It was found that a very high specific capacitance up to 276 F/g can be achieved for properly reduced GO. These rGO also show very good rate performance and cycling stability. The superior capacitive performance of rGO from the solvothermal method could be attributed to the pseudocapacitance of oxygen residual functional groups, and good dispersion and wetting property of partially reduced GO.

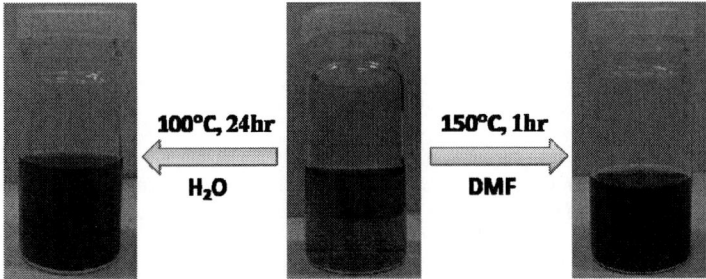

Figure 2. Pictures of GO and thermally reduced GO dispersions.

CVD Growth of Graphene

The physical properties of synthesized graphene depend on the number of layers of graphene. For example, the room temperature thermal conductivity of graphene changes from ~2800 to~1300Wm^{-1}K^{-1} when the number of layers increases from 2 to 4. Different applications may have specific requirements for the layer number of graphene. To make best use of graphene's unique properties, it is critical to control the number of layers.

We have systematically studied the growth parameters of graphene in an atmosphere CVD system, including growth substrate types, carrier gases, types of carbon sources, growth temperature, growth time, and cooling rates. It is found that single layer graphene (SLG) can be grown on a Cu substrate by introducing a small amount of a liquid hexane precursor (16). SEM image in Figure 3a and optical image Figure 3b show that the graphene is highly uniform and continuous. Raman data in Figure 3c show a sharp G' peak with a full width at half maximum of ~27 cm^{-1} and a small G/G' ratio, which are signatures of SLG. All the scanned areas show distinct features of SLG, indicating a high percentage of SLG coverage in the graphene film. Moreover, it is worth noting that no D peak was observed in the graphene film, further confirming the high quality of graphene. Another advantage of this method is that the SLG growth can be achieved at a wide temperature range. For example, graphene films grown at lower temperature down to 900 °C still show good SLG features, as seen in Figure 3c.

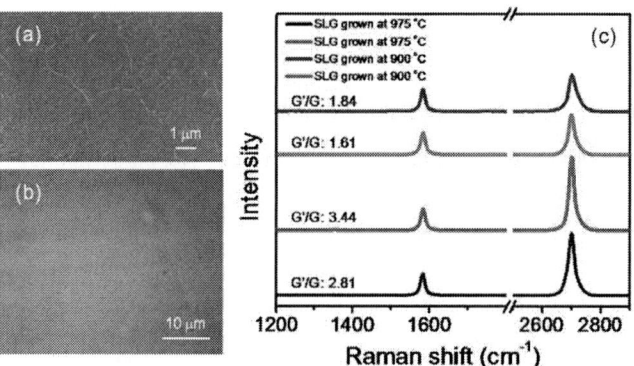

Figure 3. Typical SEM (a), and optical microscopy (b) Raman spectroscopy (c), characterization results of graphene films grown on Cu substrates using hexane as the carbon source at the growth temperature of 975 or 900 °C for 3 s. For Raman spectra, Cu substrate background was subtracted. The spectra were normalized with the G peak.

Conclusion

The practical applications of graphene highly rely on its preparation method, where low-cost, large-scale and green approaches are desirable. Toward this goal, we have developed several methods including the rapid MW reduction of GO, the low temperature solvothermal reduction of GO, and the CVD growth of graphene in an atmosphere CVD furnace. Some studies have been done to investigate the mechanisms as well as advantages of these methods for preparing supercapacitor electrode materials.

Acknowledgment

This work was supported by NSF (award # 0800849).

References

1.   K. S. Novoselov, A. K. Geim, S. V. Morozov, D. Jiang, Y. Zhang, S. V. Dubonos, I. V. Grigorieva and A. A. Firsov, *Science*, **306**, 666 (2004).
2.   K. I. Bolotin, K. J. Sikes, Z. Jiang, M. Klima, G. Fudenberg, J. Hone, P. Kim and H. L. Stormer, *Solid State Commun.*, **146**, 351 (2008).
3.   A. A. Balandin, S. Ghosh, W. Z. Bao, I. Calizo, D. Teweldebrhan, F. Miao and C. N. Lau, *Nano Lett.*, **8**, 902 (2008).
4.   R. R. Nair, P. Blake, A. N. Grigorenko, K. S. Novoselov, T. J. Booth, T. Stauber, N. M. R. Peres and A. K. Geim, *Science*, **320**, 1308 (2008).
5.   C. Lee, X. D. Wei, J. W. Kysar and J. Hone, *Science*, **321**, 385 (2008).
6.   M. D. Stoller, S. J. Park, Y. W. Zhu, J. H. An and R. S. Ruoff, *Nano Lett*, **8**, 3498 (2008).
7.   Y. Hernandez, V. Nicolosi, M. Lotya, F. M. Blighe, Z. Y. Sun, S. De, I. T. McGovern, B. Holland, M. Byrne, Y. K. Gun'ko, J. J. Boland, P. Niraj, G. Duesberg, S. Krishnamurthy, R. Goodhue, J. Hutchison, V. Scardaci, A. C. Ferrari and J. N. Coleman, *Nat. Nanotechnol.*, **3**, 563 (2008).
8.   C. Berger, Z. M. Song, T. B. Li, X. B. Li, A. Y. Ogbazghi, R. Feng, Z. T. Dai, A. N. Marchenkov, E. H. Conrad, P. N. First and W. A. de Heer, *J. Phys. Chem. B*, **108**, 19912 (2004).
9.   A. Reina, X. T. Jia, J. Ho, D. Nezich, H. B. Son, V. Bulovic, M. S. Dresselhaus and J. Kong, *Nano Lett.*, **9**, 30 (2009).
10.   S. Stankovich, D. A. Dikin, R. D. Piner, K. A. Kohlhaas, A. Kleinhammes, Y. Jia, Y. Wu, S. T. Nguyen and R. S. Ruoff, *Carbon*, **45**, 1558 (2007).
11.   H. Y. He, J. Klinowski, M. Forster and A. Lerf, *Chem. Phys. Lett.*, **287**, 53 (1998).
12.   Z. Li, Y. G. Yao, Z. Y. Lin, K. S. Moon, W. Lin and C. P. Wong, *J. Mater. Chem.*, **20**, 4781 (2010).
13.   K. Moon, Z. Li, Y. Yao, Z. Lin, Q. Liang, J. Agar, M. Song, M. Liu and C. P. Wong, *Proceedings 60th Electron. Compon. Technol. Conf. (Ectc)*, 1323 (2010).
14.   Z. Y. Lin, Y. G. Yao, Z. Li, Y. Liu, Z. Li and C.-P. Wong, *J. Phys. Chem. C*, **114**, 14819 (2010).
15.   Z. Y. Lin, Y. Liu, Y. G. Yao, O. J. Hildreth, Z. Li, K. Moon and C. P. Wong, *J Phys Chem C*, **115**, 7120 (2011).
16.   Y. G. Yao, Z. Li, Z. Y. Lin, K. S. Moon, J. Agar and C. P. Wong, *J Phys Chem C*, **115**, 5232 (2011).

Application of Graphene and Carbon Nanotubes to Transistors and Interconnects

Shintaro Sato[a,b,c], Kenjiro Hayashi[a,c], Katsunori Yagi[a], Daiyu Kondo[a,c], Ayaka Yamada[a,c], Naoki Harada[a,c], Mizuhisa Nihei[a,b,c], and Naoki Yokoyama[a,c]

[a] Green Nanoelectronics Center, AIST
16-1 Onogawa, Tsukuba, Ibaraki 305-8569, Japan
[b] MIRAI-Selete
10-1 Morinosato-Wakamiya, Atsugi, Kanagawa 243-0197, Japan
[c] Fujitsu Laboratories Ltd.
10-1 Morinosato-Wakamiya, Atsugi, Kanagawa 243-0197, Japan

We are trying to employ nano-carbon materials, such as graphene and carbon nanotubes (CNTs), as channel and interconnect materials to realize low-power-consumption large-scale integrated circuits (LSIs). In this paper, we first explain our recent progress on the application of graphene to transistor channels. Graphene synthesis on a 200-mm Si wafer by chemical vapor deposition (CVD) and electrical properties of CVD-graphene transistors are described. Especially, details of graphene growth on Cu film, such as nucleation behavior of graphene islands depending on the growth condition, are addressed. Efforts to realize CNT interconnects at MIRAI-Selete are also reviewed. Special emphases are placed on the fabrication process of CNT vertical interconnects and their reliability.

Introduction

Nano-carbon materials, such as graphene and carbon nanotubes (CNTs), are a promising candidate for future electronics devices due to excellent electrical, thermal, and mechanical properties. Graphene and CNTs are similar in properties, because CNTs are essentially rolled graphene sheets. However, each material may find different applications due to their difference in shape.

Graphene has been attracting much attention since electrical properties of mechanically exfoliated graphene were first reported in 2004 (1). Because of its unique electronic band structure around the K and K' points, electrons can behave as massless Dirac fermions, leading to interesting phenomena, such as the anomalous quantum Hall effect and Klein tunneling (2-4), which have been under intense investigation in physics. At the same time, graphene is also an attractive material in electronics mainly due to its high carrier mobility. In fact, a carrier mobility as high as 200,000 cm2/Vs has been experimentally observed (5), while a mobility of ~2,000,000 cm2/Vs has been predicted theoretically (6). Its excellent thermal and mechanical properties, as well as its planar shape, are also suitable for electronics. One possible application of graphene is field effect transistors (FETs) for large scale integrated circuits (LSIs). Due to the increasing difficulty in shrinking the device dimensions to improve its performance including the power consumption, channel materials with mobility higher than that of Si, such as Ge and GaAs, have been proposed (7). In this context, graphene with a high mobility can also be a candidate for the channel material in future LSIs. Moreover, a two-dimensional graphene channel may solve a problem called "short channel effects" (8). Furtheremore, graphene can be used as a channel for high-frequency devices, possibly replacing compound semiconductors.

CNTs are also an attractive material for a transistor channel, as demonstrate in, for example, ref. 9 and 10. However, our main interest is in their application to LSI interconnects. In fact, the development of CNT vertical interconnect (via) has been being done at Fujitsu and MIRAI-Selete for years (11-16). This is because of the fact that CNTs can sustain a current density exceeding 10^9 A/cm^2 (17, 18) and that Cu cannot sustain a high current density required for future LSIs due to electro-migration problems (7). In the approach above, a bundle of CNTs is used as a vertical interconnect. CNTs also have a high thermal conductivity of ~3,000 W/Km (19, 20). Therefore, CNT interconnects may also be used to release heat in a LSI chip (21).

In this paper, we explain our recent progress on the application of graphene to transistors, and review the efforts at MIRAI-Selete to realize CNT vias. In the first part, issues regarding the graphene transistors are addressed. Our approach to grow graphene uniformly on a large substrate by chemical vapor deposition (CVD) is described. Graphene nucleation processes are also investigated, which would help improve the quality of graphene further. A newly-developed fabrication process of graphene transistors without using a graphene-transfer process is then described (22). In the second part, the application of CNTs to LSI vias is explained, which is essentially the review of studies at MIRAI-Selete. A fabrication process of CNT vias is described in detail. Reliabilities of CNT interconnects under a high current density are then addressed.

CVD Growth of Graphene and its Application to Transistors

Graphene Growth by Chemical Vapor Deposition

Growth Condition Using Cu Catalyst Film. Graphene was grown on a Cu catalyst film deposited on a SiO$_2$/Si wafer. The thickness of Cu films was 500 nm or 1000 nm. The source gas was C$_2$H$_4$ diluted by Ar/H$_2$. The total gas pressure was kept at 1 kPa. The typical growth temperature was 860°C, which was measured with a pyrometer. The substrate was first annealed for 20 min in Ar/H$_2$ mixture (10:1), and C$_2$H$_4$ was then added for growth. The growth time and the partial pressure of the C$_2$H$_4$ were changed to optimize the growth condition. The synthesized graphene was characterized by scanning electron microscopy (SEM, Hitachi, S-4800), transmission electron microscopy (TEM; Hitachi, H-9000UHR III) and Raman spectroscopy (Jobin Yvon, LabRAM HR-800).

Graphene Synthesized on Cu In our previous study, Fe films were used as catalyst, where multi-layer graphene was mainly obtained and the reduction of the source gas supply was necessary to obtain few-layer graphene (22). On the other hand, multi-layer graphene was not obtained on Cu even when the source gas supply was relatively high, as in the report by Li et al. (23). Figures 1(a)-(c) show cross sectional TEM images of graphene for different source-gas-supply conditions. Single-layer graphene appeared to be obtained for all the conditions. The coverage of the Cu surface by graphene seems to have been affected by the source-gas supply, but we have not yet obtained the quantitative data for that. We have also found that the partial pressure of C_2H_4, P_s, itself affects the quality of graphene. Figure 1(d) shows Raman spectra for the samples shown in Figs. 1(a)-(c). We can see that the D-band intensity is high when P_s is high. This might be because the domain size of graphene is small at a higher P_s, as suggested by Li et al. (24). In order to investigate the influence of P_s further, we did SEM analyses. Figure 2 shows the results. Graphene islands grown at P_s of 0.078 Pa yielded mottled patterns (dark regions) on the Cu surface where some islands stepped over the Cu grain boundaries as can be seen in Fig. 2(a). As P_s increases, while the total amount of source gas supply [P_s × growth time] is kept constant, the island size becomes smaller and the islands' density becomes higher as shown in Fig. 2(b) (P_s = 0.59 Pa). In addition, the islands form into a scale-like shape and they tend to be aligned along steps on the Cu surface. This result indicates that the surface morphology such as grain boundaries and surface steps strongly influences the graphene growth mode under this condition. The graphene islands grown at P_s of 3.9 Pa (Fig. 2(c)) can be seen to be elongated along the steps. This kind of island shape is different from those reported in previous studies [24-26]. It was also reported that the shape of islands depended on the total gas pressure [26], but our results clearly indicate that the shape also depends on P_s.

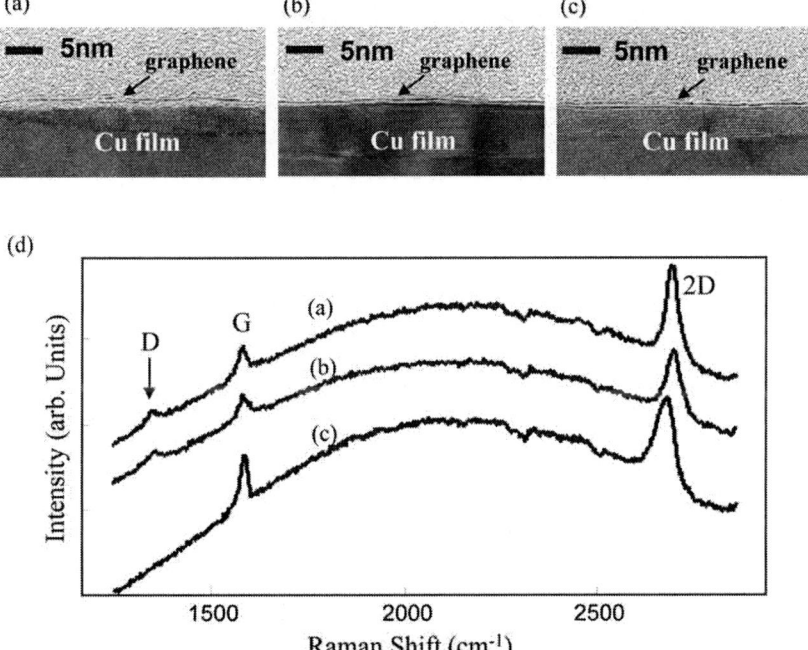

Figure 1. Cross sectional TEM images of graphene synthesized at various partial pressures of C_2H_4 (P_s) and growth time (t_g): (a) P_s = 4.7 Pa, and t_g = 1 min, (b) P_s = 4.7 Pa, and t_g = 60 min, and (c) P_s = 0.078 Pa, and t_g = 60 min. (d) Raman spectra of graphene samples shown in (a), (b), and (c).

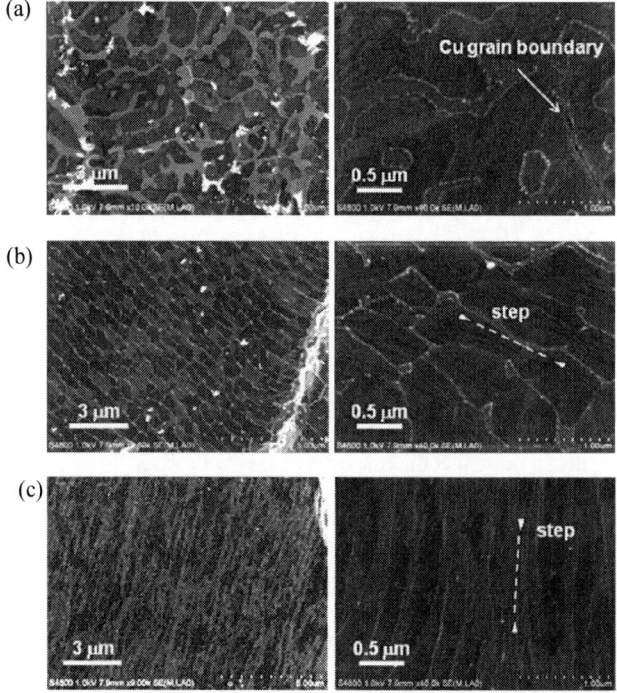

Figure 2. SEM images of graphene grown on Cu at various partial pressures of C_2H_4 (P_s) and growth time (t_g), while the product of P_s and t_g is kept constant: (a) $P_s = 0.078$ Pa, and $t_g = 30$ min, (b) $P_s = 0.59$ Pa, and $t_g = 4$ min, and (c) $P_s = 3.9$ Pa, and $t_g = 0.6$ min.

We then synthesized graphene on a 200-mm Cu (1000nm)/SiO_2/Si wafer as shown in Fig. 3(a). In this case, P_s was 0.59 Pa and growth time was 4 min. Raman spectra taken at 5 different positions on the wafer are shown in Fig. 3(b). The Raman spectra suggest that graphene was uniformly grown all over the wafer. Figure 3(c) shows a cross sectional TEM image taken around the center of the wafer. The image indicates that the monolayer graphene is formed. Incidentally, the crystalline directions of grains of a 500-nm Cu film after graphene growth were analyzed by electron back-scattered diffraction (EBSD, not shown). The average size of the grains was about 14 μm. Most of the Cu grains had the [111] direction perpendicular to the surface (99.5% of the grains were in the directions of [111] ± 10°) in contrast to the case of Cu foil where the dominant direction was [100] perpendicular to the surface (27). This suggests that the crystalline direction is not very important for graphene growth on Cu. To investigate the electrical properties of graphene, the graphene was transferred to a SiO_2/Si substrate using the conventional transfer process (28), and a back gate transistor was fabricated. Figure 3(d) shows the drain current as a function of the gate voltage. The values of field effect mobility were around 1,700 cm^2/Vs and 2,000 cm^2/Vs before and after subtracting the contact resistance, which are of the same order as those in the recent study (24).

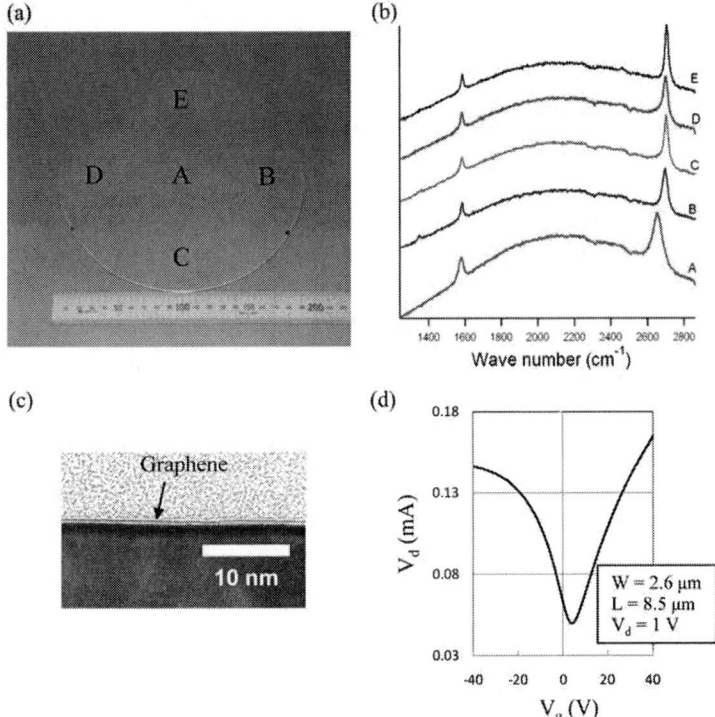

Figure 3. (a) Graphene synthesized on a 200-mm Cu/SiO$_2$/Si wafer. (b) Raman spectra of graphene at positions of A-E shown in (a). (c) A cross sectional TEM image of the graphene around the center of the wafer. (d) Drain current as a function of the gate voltage for a back-gated graphene transistor (channel width (W): 2.6 µm, channel length (L): 8.5 µm, drain voltage (V$_d$): 1 V, gate-oxide thickness: 300 nm).

Fabrication of Graphene-Channel Top-Gated Transistors Using a Transfer-Free Process

Transfer-Free Fabrication Process The fabrication process of a top-gated graphene FET is schematically described in Fig. 4 (22). In this example, graphene grown on Fe film at 650°C was used. The detailed growth condition is explained elsewhere (22). First, Fe films with channel patterns were prepared on a SiO$_2$/Si substrate by conventional photolithography and lift-off processes. Graphene was then synthesized on the patterned Fe films by thermal CVD. After synthesis, source and drain electrodes consisting of Au (300 or 50 nm)/Ti (10 nm) were formed by electron-beam evaporation at both ends of the patterned graphene, under which the Fe films remained. The Fe films were then removed by wet etching using dilute HCl, leaving suspended or partially suspended graphene channels bridging the source and drain electrodes. The graphene channels were subsequently covered by a HfO$_2$ layer with a typical thickness of 70 nm by atomic layer deposition (ALD). The ALD temperature was 250°C, and the precursors were tetrakis (dimethyl) amino hafnium (TDMAH) and H$_2$O. Finally, top-gate electrodes of Au (50 nm)/Ti (10 nm) were formed. The length and width of graphene channels were typically 1–5 μm and 2–10μm, respectively. Incidentally, fabrication of top-gated transistors using CVD graphene by transfer-free processes was also reported by another research group (29). However, in that study, the electrode material was the same as the catalyst material, causing the electrodes to be etched during the catalyst removal. In contrast, our method uses electrode materials different from the catalyst, resulting in better etching selectivity and compatibility with smaller-sized device fabrication.

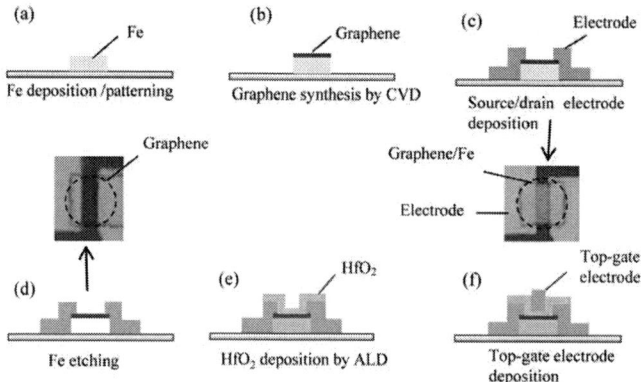

Figure 4.   (a)-(f) Fabrication process of a top-gated graphene FET. The optical microscopy images of the device before and after etching the Fe film are also shown.

Appearance and Performance of Transistors Graphene transistors were fabricated on substrates as large as 75 mm, as shown in Fig. 5(a), but our process can be applied to much larger substrates. An optical microscope image of a top-gated transistor is shown in Fig. 5(b). A cross-sectional TEM image of a transistor, together with a schematic diagram of the cross section of the device, is shown in Fig. 5(c). It can be seen that the graphene channel is suspended with HfO_2 films deposited on both sides, which is due to the conformal nature of ALD.

Measurements of the top-gated FETs were done in air at room temperature. A typical result for the drain current as a function of the top-gate voltage is shown in Fig. 5(d). The transconductance of this device is 10 mS/mm. Since the dielectric constant of the HfO_2 film was 16, the field effect mobility is estimated to be 110 cm^2/Vs. Our mobility calculation is based on the current including the contact resistance, so the intrinsic mobility of the channel should be higher than this value. We also obtained devices with a transconductance as high as 22 mS/mm and a field-effect mobility of 230 cm^2/Vs. The filed-effect mobility is actually not as good as that of top-gated FETs using exfoliated graphene by Farmer et al. (30). It is also lower than that of the back gate transistor describe in the previous section. The lower mobility might be partly because the quality of the graphene is not very good due to the low growth temperature. In addition, the HfO_2 deposition processes might have damaged the graphene channels. In order to investigate the variations of graphene-channels, the sheet resistivities of 38 graphene channels were measured. The results are shown in Fig. 5(e). By fitting a log-normal distribution, the geometric mean was estimated to be 760 Ω/sq., with a geometric standard deviation of

2.2. The variations may have been caused by those in the number of layers and/or those in the contact resistance. Incidentally, the current density in the graphene channel was 10^7–10^8 A/cm^2 for the device shown in Fig. 5(d), assuming a graphene thickness of ~1 nm. This suggests that graphene can also be a robust wiring material for future LSIs.

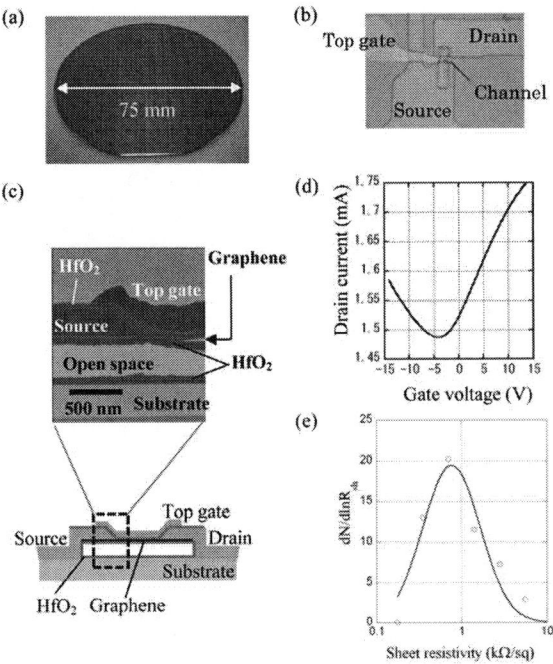

Figure 5. (a) A 75-mm wafer on which graphene-channel FETs were fabricated. (b) An optical microscopy image of a top-gated FET. (c) A cross-sectional TEM image of a device and its schematic diagram. (d) Drain current as a function of the top-gate voltage (channel width: 2 μm, channel length: 3 μm, drain voltage: 1.4 V, HfO$_2$ thickness: 70 nm). (e) Distribution of sheet resistivity (R$_{sh}$) of graphene channels including the contact resistance. The vertical axis shows the number normalized by the logarithmic range width. The fitting curve for a log-normal distribution is also shown.

Carbon Nanotube Interconnects

Fabrication process of carbon nanotube interconnect

A typical fabrication process of a CNT via is shown schematically in Fig. 6 (31). A substrate with a horizontal Cu interconnect covered by a dielectric layer was first prepared. The dielectric layer was SiOC with a dielectric constant (k) of 3.0 or 2.6. Via holes with a diameter of 160 nm were made using the conventional photolithography followed by dry etching. A TaN/Ta barrier layer and a TiN contact layer were deposited by physical vapor deposition (PVD). Size-controlled Co particles with a mean diameter of about 4 nm were then deposited using a custom-designed particle generation and deposition system, which is explained elsewhere (15, 32). Multi-walled CNTs (MWNTs) were then grown by thermal CVD using C_2H_2 diluted by Ar as the source gas. The pressure of the source gas was kept at 1 kPa. The growth temperature ranged from 400°C to 450°C. The substrate with MWNTs was then coated with spin-on glass (SOG) and planarized by chemical mechanical polishing (CMP). The CMP condition was similar to that used for polishing a silicon dioxide layer. Actually, the substrate was polished with the conventional IC1000 pad and silica slurry at a pressure of 2 psi (13.8 kPa) for 150 sec. Finally, a Ti top contact layer, a Ta barrier layer and a Cu wire were deposited on the CNT vias by PVD.

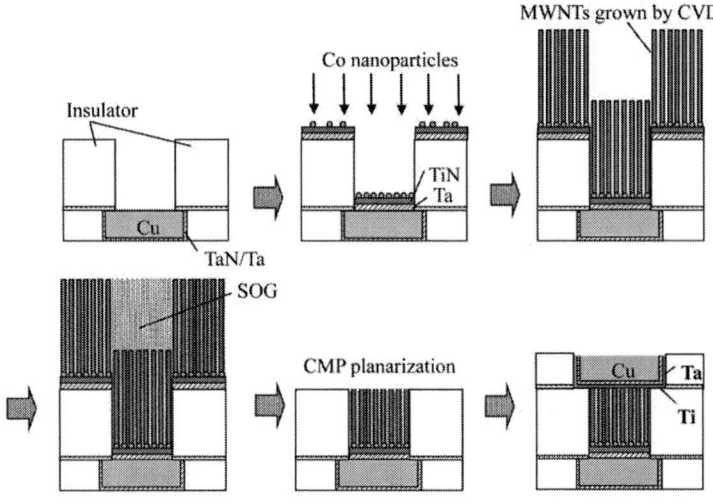

Figure 6. Typical fabrication process of a CNT via.

Carbon Nanotube Synthesized at Low Temperature

Figures 7(a) and 7(b) are cross-sectional SEM images of MWNTs grown in via holes with a diameter of 160 nm at growth temperatures of 450°C and 400°C (14). It can be seen that MWNTs grown at 400°C are a little less straight than those grown at 450°C, suggesting that MWNTs at 400 °C are a little more defective. To investigate the quality of MWNTs, TEM observation was performed, and the results are shown in Figs. 7(c) and 7(d). The TEM images indicate that MWNTs grown at either temperature definitely have a tubular structure, although, again, MWNTs at 400 °C look a little more defective. Incidentally, it should be noted that the low-temperature growth achieved is partly due to the existence of the TiN layer under Co particles. In fact, CNTs could not be grown under the same growth conditions without the TiN layer. This result is similar to former results showing that alloy particles consisting of Co and Ti are more effective as a catalyst for CNT growth than pure Co particles (33).

The average site density of MWNTs in the 160-nm via holes was 3×10^{11} cm^{-2}, while that of Co particles was about 5×10^{12} cm^{-2}, suggesting that the growth probability of

MWNTs is still low. However, recently, the growth of higher-density CNTs has been achieved, as shown in Fig. 7(e) (34). The site density is estimated to be as high as 10^{12} cm^{-2}. It should be noted, however, that the electrical properties explained in the next section were measured for vias consisting of CNTs with a site density of 3×10^{11} cm^{-2}.

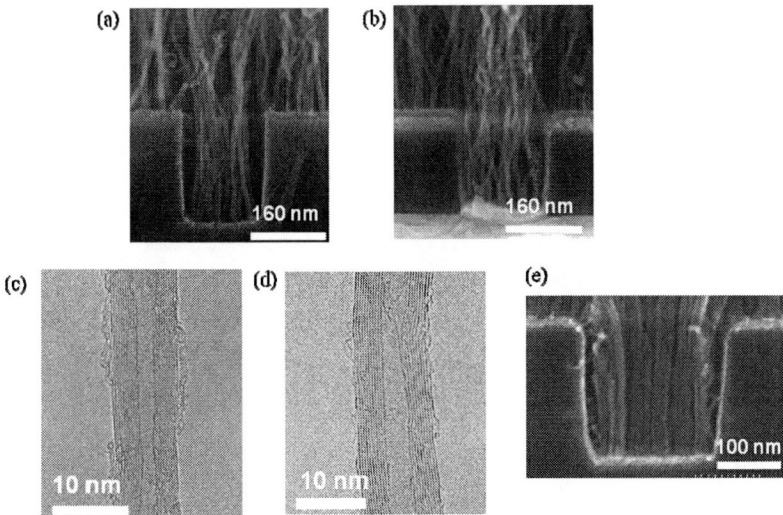

Figure 7. SEM images of MWNTs grown at (a) 450 °C and (b) 400 °C, and TEM images of MWNTs grown at (c) 450 °C and (d) 400 °C. (e) SEM image of MWNTs with a density of ~10^{12} cm^{-2} grown at 450 °C.

Electrical Properties and Reliability of CNT vias

The resistances of 160-nm CNT vias were measured with a four-point probe using Kelvin patterns. The via height was 120 nm. The current-voltage characteristics are shown in Fig. 8(a) (14). It was found that the resistance depended on the growth temperature. The via resistance was 34 Ω for the growth temperature of 450 °C, and 64 Ω for 400 °C. Since the site density of the CNTs was about 3.0×10^{11} cm^{-2} for both the temperatures, it is considered that the difference in resistance was caused by the difference in CNT quality. The resistance obtained at 450 °C is of the same order as that of W plugs.

The stability of the via resistance under an electric current with a density of 5.0×10^6 A/cm^2 is shown in Fig. 8(b) (14). The via diameter and growth temperature were 160 nm and 400°C, respectively. The dielectric layer was made of SiOC with k = 2.6 (referred to as "ULK"). The measurement was performed at 105 °C in vacuum. The resistance remained stable even after running the electric current for 100 hours, indicating that the CNT via is robust over a high-density current. The robustness of CNT vias demonstrated here, however, is not necessarily better than that of Cu. This is mainly because the MWNTs filled only part of the space of the via hole. It is expected that CNT vias with a higher CNT density would be much more robust than Cu vias. Incidentally, the via shape appears deformed in the TEM image shown in Fig. 8(c), but this was actually caused by irradiation of high-energy electrons during the TEM observation. Although the ULK is vulnerable to heat, it was found that it was not damaged during CNT growth due to the low growth temperature.

Reliability tests were further performed while changing the interface structures between CNTs and the electrodes. It has been found that the interface structures greatly affect the reliability of CNT vias. The detailed results are shown elsewhere (35). Here suffice it to say that, by improving the contacts between the CNTs and the electrode, CNT vias become much more reliable, leading to failures of horizontal Cu interconnects before those of CNT vias.

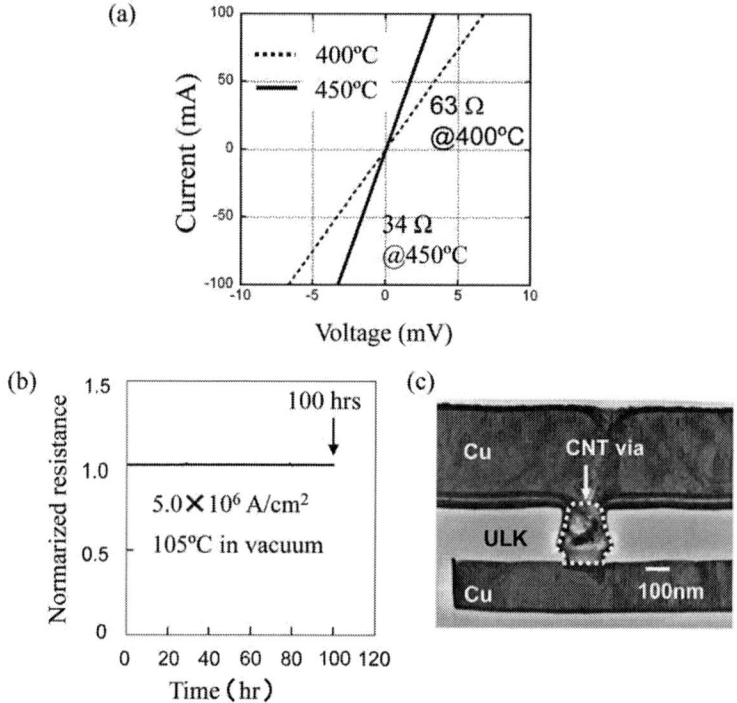

Figure 8. (a) Current-voltage characteristics of 160-nm CNT vias for growth temperatures of 400 °C and 450°C. (b) Stability of the via resistance under an electric current of a density of 5.0×10^6 A/cm^2. (c) Cross sectional TEM image of the CNT via.

Conclusions

Graphene and carbon nanotubes are promising candidates for materials for future LSIs. In this paper, we explain the application of graphene and CNTs to a transistor channel and interconnect, respectively. For the application of graphene to transistors, we demonstrated uniform graphene growth all over a 200 mm wafer. The initial stages of graphene growth were also investigated in order to improve the quality of graphene. We also developed the transfer-free fabrication process of top-gated FETs utilizing CVD-synthesized graphene. Although there are still many issues to address for realizing graphene transistors, such as band-gap control, our study shown here is an important step toward the future goal. As for CNT interconnects, the efforts at MIRAI-Selete were reviewed. The low-temperature CNT growth and the fabrication process of CNT vias were presented. In addition, the electrical properties and reliabilities of CNT vias were briefly explained. The electrical properties shown here are not yet good enough to replace Cu vias, mainly due to the insufficient site density of CNTs. However, the recent improvement in the site density is expected to lead to better electrical properties, replacing Cu with CNTs eventually.

Acknowledgments

The graphene-related work was partly supported by the Japan Society for the Promotion of Science (JSPS) through its "Funding Program for World-Leading Innovative R&D on Science and Technology (FIRST Program)." The work related to CNT interconnects was completed as part of the MIRAI Project supported by NEDO. We would like to thank Dr. Mogami and Prof. Awano for their initiative for the CNT-interconnect program at MIRAI-Selete.

References

1. K. S. Novoselov, A. K. Geim, S. V. Morozov, D. Jiang, Y. Zhang, S. V. Dubonos, I. V. Grigorieva, and A. A. Firsov: *Science*, **306**, 666 (2004).
2. K. S. Novoselov, A. K. Geim, S. V. Morozov, D. Jiang, M. I. Katsnelson, I. V. Grigorieva, S. V. Dubonos, and A. A. Firsov: *Nature*, **438**, 197 (2005).
3. Y. Zhang, Y.-W. Tan, H. L. Stormer, and P. Kim: *Nature*, **438**, 201 (2005).
4. M. I. Katsnelson, K. S. Novoselov, and A. K. Geim: *Nature Phys.* **2**, 620 (2006).
5. K. I. Bolotin, K. J. Sikes, Z. Jiang, M. Klima, G. Fudenberg, J. Hone, P. Kim, and H. L. Stormer, *Solid State Commun.*, **146**, 351 (2008).
6. E. H. Hwang, S. Adam, and S. Das Sarma: *Phys. Rev. Lett.*, **98**, 186806 (2007).
7. International Technology Roadmap for Semiconductors (http://www.itrs.net/home.html)
8. S. M. Sze, *SEMICONDUCTOR DEVICES: Physics and Technology, 2 nd Edition*, John Wiley and Sons, Inc., Hoboken, NJ (2002).
9. Z. Zhang, S. Wang, Z. Wang, L. Ding, T. Pei, Z. Hu, X. Liang, Q. Chen, Y. Li, and L. Peng: *ACS Nano*, **3**, 3781 (2009)
10. N. Moriyama, Y. Ohno K. Suzuki, S. Kishimoto, and T. Mizutani: *Appl. Phys. Express*, **3**, 105102 (2010)
11. M. Nihei, M. Horibe, A. Kawabata, and Y. Awano: *Proc. IEEE Int. Interconnect Technology Conf. 2004*, p. 251 (2004).
12. M. Nihei A. Kawabata, D. Kondo, M. Horibe, S. Sato, Y. Awano: *Jpn. J. Appl. Phys.* **44** (2005) 1626.
13. M. Horibe, M. Nihei, D. Kondo, A. Kawabata and Y. Awano: *Jpn. J. Appl. Phys.*, **44**, 5309 (2005).

14. A. Kawabata, S. Sato, T. Nozue, T. Hyakushima, M. Norimatsu, M. Mishima, T. Murakami, D. Kondo, K. Asano, M. Ohfuti, H. Kawarada,T. Sakai, M. Nihei, Y. Awano: *Proc. IEEE Int. Interconnect Technology Conf. 2008*, p. 237 (2008).
15. S. Sato, A. Kawabata, T. Nozue, D. Kondo, T. Murakami, T. Hyakushima, M. Nihei, and Y. Awano: *Sensors and Materials*, **21**, 373 (2009).
16. Y. Awano, S. Sato, M. Nihei, T. Sakai, Y. Ohno, and T. Mizutani: *Proc. IEEE*, **98**, 2015 (2010).
17. B. Q. Wei, R. Spolenak, Ph. Kohler-Redlich, M. Ru¨hle, and E. Arzt: *Appl. Phys. Lett.*, **74**, 3149 (1999).
18. B. Q. Wei, R. Vajtai, and P. M. Ajayan: *Appl. Phys. Lett.*, **79**, 1172 (2001).
19. J. Hone, M. Whitney, C. Piskoti, and A. Zettl: *Phys. Rev. B*, **59**, R2514 (1999).
20. P. Kim, L. Shi, A. Majumdar, and P. L. McEuen: *Phys. Rev. Lett.* **87**, 215502 (2001).
21. N. Srivastava, R. V. Joshi, and K. Banerjee: *IEDM Tech. Dig. 2005*, p. 249 (2005).
22. D. Kondo, S. Sato, K. Yagi, N. Harada, M. Sato, M. Nihei, and N. Yokoyama, *Appl. Phys. Express*, **3**, 025102 (2010).
23. X. Li, W. Cai, J. An, S. Kim, J. Nah, J. Dang, R. Piner, A. Velamakanni, I. Jung, E. Tutuc, S. K. Banerjee, L. Colombo, and R. S. Ruoff, *Science*, **324**, 1312 (2009).
24. X. Li, C. W. Magnuson, A. Venugopal, J. An, J. W. Suk, B. Han, M. Borysiak, W. Cai, A. Velamakanni, Y. Zhu, L. Fu, E. M. Vogel, E. Voelkl, L. Colombo, and Rodney S. Ruoff, *Nano Lett.*, **10**, 4328 (2010).
25. A. W. Robertson and J. H. Warner: *Nano Lett.*, **11**, 1182 (2011).
26. Qingkai Yu, L. A. Jauregui, W. Wu, R. Colby, J. Tian, Z. Su, H. Cao, Z. Liu, D. Pandey, D. Wei, T. F. Chung, P. Peng, N. Guisinger, E. A. Stach, J. Bao, S-S. Pei, Y. P. Chen, *arXiv:*1011.4690v1 (2010).
27. J. M. Wofford, S. Nie, K. F. McCarty, N. C. Bartelt, and O. D. Dubon, *Nano Lett.*, **10**, 4890 (2010).
28. A. Reina, X. Jia, J. Ho, D. Nezich, H. Son, V. Bulovic, M. S. Dresselhaus, and J. Kong, *Nano Lett.*, **9**, 30 (2009).
29. M. P. Levendorf, C. S. Ruiz-Vargas, S. Garg, and J. Park, *Nano Lett.* **9**, 4479 (2009).
30. D. B. Farmer, H.-Y. Chiu, Y.-M. Lin, K. A. Jenkins, F. Xia, and Ph. Avouris, *Nano Lett.*, **9**, 4744 (2009).
31. M. Nihei, T. Hyakushima, S. Sato, T. Nozue, M. Norimatsu, M. Mishima, T. Murakami, D. Kondo, A. Kawabata, M. Ohfuti, Y. Awano: *Proc. IEEE Int. Interconnect Technology Conf. 2007*, p. 204 (2007).
32. S. Sato, M. Nihei, A. Mimura, A. Kawabata, D. Kondo, H. Shioya, T. Iwai, M. Mishima, M. Ohfuti, and Y. Awano: *Proc. IEEE Int. Interconnect Technology Conf. 2006*, p. 230 (2006).
33. S. Sato, A. Kawabata, D. Kondo, M. Nihei, and Y. Awano: *Chem. Phys. Lett.*, **402**, 149 (2005).
34. Y. Awano: *Proc. Selete Symp. 2009*, p. 115 (2009).
35. M. Sato, T. Hyakushima, A. Kawabata, T. Nozue, S. Sato, M. Nihei, and Y. Awano: *Jpn. J. Appl. Phys.*, **49**, 1051102 (2010).

ECS Transactions, 37 (1) 133-139 (2011)
10.1149/1.3600733 ©The Electrochemical Society

Manipulation of Graphene Properties by Interface Engineering

X. M. Wang[a], J. B. Xu[a], C. L. Wang[a], W. G. Xie[a], and J. Du[a]

[a] Department of Electronic Engineering, and Materials Science and Technology Research
Center, The Chinese University of Hong Kong, Shatin, N. T., Hong Kong SAR,China
E-mail: jbxu@ee.cuhk.edu.hk

Graphene is generally regarded as an ideal candidate material for
post-CMOS nanoelectronics. In contrast to traditional
semiconductors, the unique two dimensional structure of graphene
offers the necessity and possibility of studying the interface
characteristics for its sensitivity to the top surface and interface
between graphene and the substrate. In this paper, we provide
preliminary studies in understanding graphene surface and
interface issues in electronic structure, carrier transport and related
phenomena down to nano-scale.

Introduction

Graphene, a new allotrope of carbon, received a great deal of interest although it was
discovered very recently (1). In contrast to traditional semiconductors, the unique two
dimensional structure of graphene offers the necessity and possibility in studying the
interface characteristics for its sensitivity to the top surface and interface between
graphene and the substrate. We are thus interested in understanding graphene surface and
interface issues in electronic structure, carrier transport and related phenomena down to
nano-scale. In this paper, we investigate the mechanisms of graphene interface coupling
to different substrates and interacting with inert adsorbates, both experimentally and
theoretically. At first, the crucial roles of the substrate (interfacial effects) played in
graphene applications are meticulously interrogated. To suppress the deleterious substrate
effect on graphene intrinsic electronic structure and carrier transport properties, we have
modified the substrate by highly ordered SAMs. The charge transfer is unambiguously
corroborated by KPFM and transport measurement results. The measured temperature
dependence of resistivity provides the evident correlation between the transport
characteristics and the phonon-electron scattering by the interplay between graphene and
the substrate. After diminishing the unwanted scattering origins, a nearly one order of
increase in mobility is obtained. Subsequently, through molecular self-assembly above
graphene by organic molecules (surface effect), both n-type and p-type doped
mechanically exfoliated graphene sheets are accomplished. By exploiting the Kelvin
probe force microscopy (KPFM), the charge transfer process identified as the key issue is
quantitatively analyzed by a self-consistent tight binding model. By combining the two
techniques, graphene PN junction will be demonstrated if p-type and n-type regions are
fabricated in a graphene sheet on the modified substrate. Moreover, the junction interface
can be monitored by KPFM. The insightful understanding of graphene PN junction
provides numerous opportunities in both novel electric and optoelectronic devices.

133

Substrate Modification

Graphene field-effect transistors (FETs) have been considered as promising devices due to graphene's notably high mobility, flexiblility and ultrathin nature. Indeed, high mobility up to 200,000 cm2/Vs and large mean free length are observed in clean suspended graphene at 5K (2). Unfortunately, the applications of graphene are usually confined by standard SiO2 substrate through charge transfer, adsorbates and electron phonon scattering (3). To circumvent this problem, traditional silicon wafers were modified by self-assembly monolayers (SAMs) (4-5). Recently, in a previous work, using SiO_2/Si substrate modified by octadecyltrimethoxysilane (OTMS, $CH_3(CH_2)_{17}Si(OCH_3)_3$) SAMs, we obtained high-quality graphene devices with low intrinsic doping level. The high mobility could reach up to 47,000 cm2/V-s (6).

Both the results verified SAMs modification is an useful tool to reduce charge transfer between substrate and graphene. To verify the negligible charge transfer between graphene and modified substrate, KPFM of graphene on OTMS modified SiO_2/Si was conducted in comparison with that on the untreated SiO_2/Si. Typical KPFM images are demonstrated in Figure 1. From the previous studies, it is found that for graphene on the untreated SiO_2/Si charges are naturally transferred from graphene to electronic traps (mainly hydroxyl groups, e.g., water and SiOH) on SiO2, which build up an electric field under ambient conditions. Moreover, the charge transfer leads to a vertical gradient in charge distribution. As reported recently, the distribution and induced field decrease the energy difference between the vacuum level and the Fermi level of few-layer graphene (FLG) monotonically, as the doping level increases (7). This phenomenon can be observed through spatial variation of contact potential difference (CPD) in Figure 1. The CPD of graphene on the untreated SiO_2 substrate increases with the layer thickness and reaches saturation, at which the electric field is adequately screened. On the contrary, the CPD of the OTMS SAM functionalized substrate is almost homogenous, indicating that there is little band bending or band mismatch between the graphene layers close to and away from the substrate, and very limited charge transfer has occurred.

Currently, contact angle is widely used to macroscopically characterize hydrophobic quality among different SAMs. However, our experiments suggest that in graphene application, the SAM microscopic structure is more important. Besides highly ordered OTMS, the surface modification by a series of low coverage density SAMs such as oxydiphthalic dianhydride (ODPA) and amorphous OTMS is also interrogated. For comparision, we fabricated GFET in typical SAMs. Although similar contact angles as large as 110° can be obtained in some materials, the doping level and room temperature mobility are improved limitedly (see Table 1).

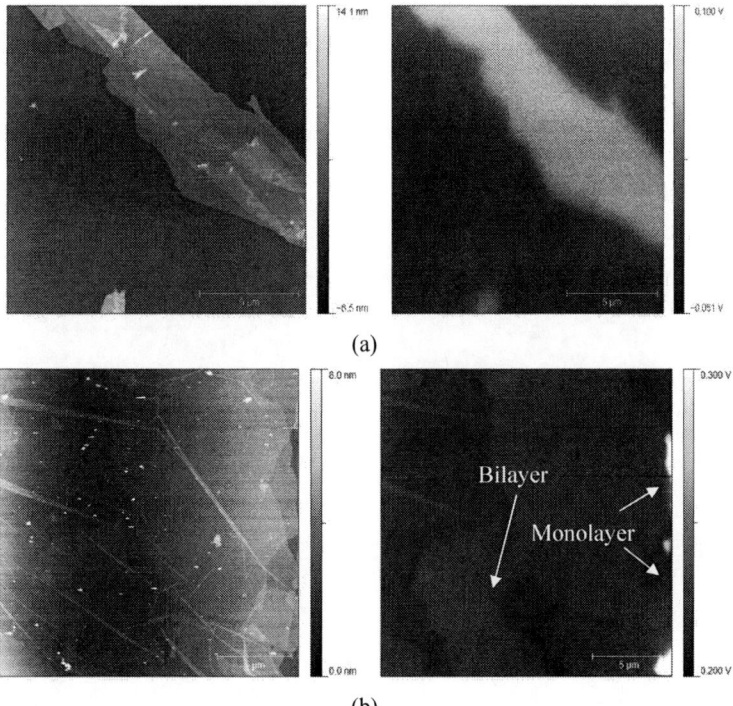

(a)

(b)

Figure 1 AFM (left column) and KPFM (right column) images of few-layer graphene (FLG) on OTMS modified SiO_2/Si and bare SiO_2/Si substrates: (a) as-prepared FLG on OTMS modified SiO_2/Si substrate; (b) as-prepared FLG on bare SiO_2/Si substrate. Scale bars are 5μm

We attribute the performance variation to the coverage of SAMs. Despite other SAMs also present good contact angle, the relatively low coverage may increase the scattering. Our results are helpful to protect the intrinsic properties of supported graphene, reveal the mystery of scattering sources and understand the mechanism of the charge transport.

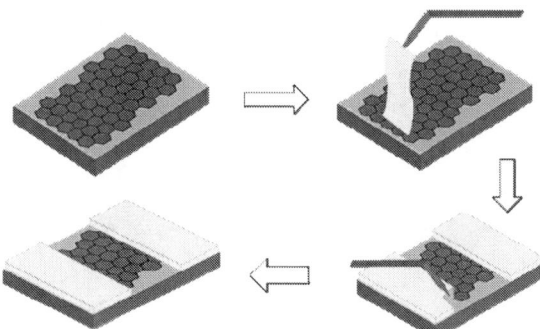

Figure 2 GFET fabrication process. The electrode is formed by tailoring 100 nm-thick Au film predeposited on another substrate. Subsequently, the electrode is tipped up by a micro manipulator, aligned to the graphene and allocated under a 400 × optical microscope. The channel width defined by this approach is typically around 10 microns.

TABLE I. Performance of GFETs on modified SiO_2/Si substrates

Substrate	Contact Angle (Degree)	Mobility
Highly-ordered OTMS	114	1.2×10^4 cm^2/ V•s
Amorphous OTMS	112	8.0×10^3 cm^2/ V•s
OTS	107	4.2×10^3 cm^2/ V•s
ODPA	104	5.0×10^3 cm^2/ V•s
Bare SiO_2	46	3.5×10^3 cm^2/ V•s

Organic Charge Transfer Doping

Atomic doping is a widely used technique in conventional semiconductor industry, which has been successfully used to achieve the n-type doping effect by controlling the concentration of potassium on ultrathin epitaxial graphene (EG) and a small band-gap between valence and conduction bands could be tuned (8). In general, alkali metals are effective dopants for n-type doping. However, the unstable chemical properties, structural deformation and strong scattering by doping species adversely affect their applications. An alternative way is to use transition metals (TM) or noble metals. By depositing gold atoms onto graphene, the p type doping effect has been realized and the Dirac point is shifted into the unoccupied states (9). The selective n-type or p-type doping could also be achieved by graphene adsorption on transitional metal surface such as Al, Ag, Cu, Au, and Pt, and the doping effect has been investigated by *ab initio* calculation (10). However, a fundamental hurdle is that the edge structures and chemical terminations of graphene synthesized by various methods are unknown and uncontrollable. Therefore the metallic doping may be perplexing.

Surface molecule functionalization is another candidate for tuning the electronic structure of graphene materials. NO_2 was used to dope single-layer or bilayer graphene sheets, and metal to insulator transition was found by angle-resolved photoemission spectroscopy (ARPES) (11). Fine-tuning of the charge carriers from electrons to holes can be achieved in the graphene sheets and the NO_2 doping effect could also be reversible. This implicates that organic molecule surface functionalization could introduce functional groups to surface to achieve intended properties. Tetracyanoethylene (TCNE) on graphene was studied by first-principles method and the band-gap opening and p-type doping effect could be tuned by the coverage of molecule dose (12). Meanwhile, nitrophenyl group was attached to graphene by covalent carbon-carbon bond and the electronic structure and transport properties of graphene were tuned from semimetal to p-type semiconductor (13). Meanwhile, Wee *et al.* have used tetrafluoro-tetracyanoquinodimethane (F4-TCNQ) to functionalize graphene and ARPES has proved the p-type doping effect of graphene by F4-TCNQ (14).

Using molecules to functionalize the graphene surface, no defect or strong structural conformation will be created and graphene remains its perfect two dimensional honeycomb structure merit. The charged molecule anions will repel each other at high coverage, so no clustering of molecular dopant will appear. By employing Kelvin probe force microscopy, we have shown that the 2,3,5,6-tetrafluoro- 7,7,8,8-tetracyanoquinodimethane (F4-TCNQ) molecules obtain electrons from graphene, whereas vanadyl-phthalocyanine (VOPc) molecules donate electrons to it (15). In additional, combined with a tight-binding model, the charge transfer can be estimated quantitatively.

Here, use the similar method, we illustrate hydrogen and petacene are efficient n-type and p-type dopant respectively. By depositing molecules on graphene, high level doping effects were obtained, as shown in figure 3.

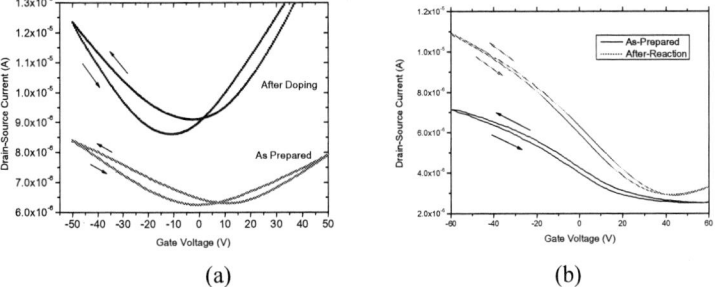

(a)   (b)

Figure 3 Transfer character of functionalize graphene (a) n-type doping by hydrogen (b) p-type doping by pentacene

Graphene PN Junction

A *p-n* junction is a basic building block in modern microelectronics. However, among graphene electron devices, in contrast to typical *p-n* junctions in conventional semicondcutor technology, the detailed properties of graphene *p-n* junction by spatially selective doping have scarcely explored, although electrically induced *p-n* junctions are recently interrogated by several groups (16). They found that the electrostatic modification of the gate insulator surface can be exploited to produce an abrupt *p-n* junction in monolayer graphene device structures, which may offer opportunities extending beyond CMOS technology. Obviously, this technique has limited applications due to the complicated electrostatic arrangement. Graphene *p-n* junctions fabricated by spatial chemical doping are highly desirable in terms of fabrication feasibility and variability.

To form a *p-n* junction by spatial chemical doping, we use a chemical doping with a sharp gradient in a predefined area or a layer step on graphene sheet to complete the task. For instance, we can place a monolayer or bilayer graphene on SiO_2/Si substrate which is halfly modified by SAM or molecular doping, as shown in Fig. 4 (a). The right half exposed directly to the substrate is naturally *p*-type doped, while the left half is protected by a SAM layer, and may be subjec to further doping by molecular materials, e.g., pentacene.

Preliminary *I-V* transfer curve of graphene *p-n* junction that we fabricated by spatial chemical doping with a device structure in Fig. 4(a) is shown in Fig. 4(b). The two dips corresponding to the Dirac points indicate the existence of two inhomogeneous channels of GFETs. The separation of the Dirac points manifests the carrier density difference and the dissimilar polarity. When the gate voltage is located between the two dips, one of the GFETs possesses electron-like carriers in the channel and the other provides a hole-like conduction. Consequently, a *p-n* junction is formed.

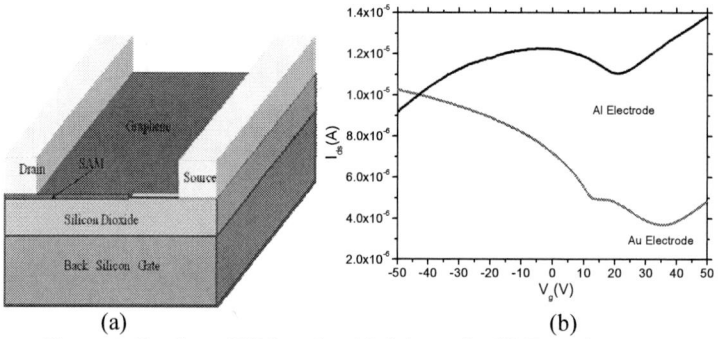

(a) (b)

Figure 4 Graphene PN Junction (a) Schematic (b) Transfer curve

Uptil now, there is no comparision of electronic and optical properties between electrostatic doped *p-n* junction and spatially chemically doped *p-n* junction, particularly, in terms of external *I-V* characteristics, internal electrical potential distribution in the *p-n* transition region, and doping inhomogeneity, etc. Intuitively, it is speculated that there would exist a marked disparity because of the different doping origins and interactions with the substrate. As a result, it requires more efforts to clarify these basic characteristics between spatially chemically doped and electrostatic doped *p-n* junctions. Besides, from an application perspective, effects of imperfections, such as graded abruptness and/or interdiffusion in *p-n* space-charge region, are also open to be explored.

Acknowledgments

The authors would like to acknowledge the colleagues in Solid-State Electronics Group, Dept. of Electronic Engineering, The Chinese University of Hong Kong, for various technical supports, especially, Dr. N. Ke for assistance in FET fabrication, Dr. X. Q. Tian and Mr. K. Chen for support in first-principles calculation, and Dr. X. D. Ding for assistance in KPFM. The work is in part supported by Research Grants Council of Hong Kong, particularly via Grants nos. CUHK2/CRF/08, CUHK4179/10E, and CUHK4182/09E. J. B. Xu thanks the National Science Foundation of China (Grants Nos. 60990314 and 60928009) for support.

References

1. A. K. Geim, K. S. Novoselov, *Nature Materials*, **6**, 183(2006).
2. K. I. Bolotin, K. J. Sikes, J. Hone, H. L. Stormer, P. Kim, *Physical Review Letters* **101**, 096802(2008).
3. S. Adam, E. W. Hwang, V. M. Galitski, S. Das Sarma, *Proceedings of the National Academy of Sciences* **104**, 18392(2007).
4. M. Lafkioti, B. Krauss, T. Lohmann, U. Zschieschang, H. Klauk, K. von Klitzing, J. H. Smet, *Nano Letters* **10**, 1149(2010).
5. Z. Liu, A. A. Bol, W. Haensch, *Nano Letters* **11**, 523(2011).
6. Xiaomu Wang, Jian-Bin Xu, Chengliang Wang, Jun Du, Weiguang Xie, *Advanced Materials*, online, DOI: 10.1002/ adma.201100476 (2011)
7. S. S. Datta, D. R. Stranchan, E. J. Mele, A. T. C. Johnson, *Nano Letters* **9**, 7-11 (2009).
8. T. Ohta, A. Bostwick, T. Seyller, K. Horn, and E. Rotenberg, *Science* **313**, 951 (2006)
9. I. Gierz, C. Riedl, U. Starke, C. R. Ast and K. Kern, *Nano Lett.* **8**, 4603, (2008)
10. G. Giovannetti, P. A. Khomyakov, G. Brocks, V. M. Karpan, J. V. D. Brink, and P. J. Kelly, *Phys. Rev. Lett.* **101**, 026803 (2008)
11. S. Y. Zhou, D. A. Siegel, A.V. Fedorov, and A. Lanzara, *Phys. Rev. Lett.* **101**, 086402 (2008).
12. Y. H. Lu, W. Chen and Y. P. Feng, P. M. He, *J. Phys. Chem. B*. **113**, 2 (2009)
13. E. Bekyarova, M. E. Itkis, P. Ramesh, C. Berger, M. Sprinkle, W. A. de Heer, and R. C. Haddon, *J. Am. Chem. Soc.*, **131**, 1336 (2009)
14. W. Chen, S. Chen, D. C. Qi, X. Y. Gao, and A. T. S. Wee, *J. Am. Chem. Soc.* **129**, 10418 (2007)
15. Xiaomu Wang, Jian-Bin Xu, Weiguang Xie, and Jun Du, *Journal of Physical Chemistry C* **115** (15), pp 7596–7602 (2011)
16. H. -Y. Chiu, V. Perebeinos, Y. -M. Lin, and Ph. Avouris, Nano Lett., **10**,4634 (2010)

140

Silicon Grow-in-Place Nanowires and Their Applications

Jian Wu, Pranav Garg, Suxing Pan, Chris Winter, Dana Scott, and Stephen Fonash

Center for Nanotechnology Education and Utilization

The Pennsylvania State University, University Park, PA 16802, USA

Grow-in-place VLS Silicon nanowires have been used since 2007 to produce accumulation metal oxide semiconductor accumulation field effect transistors. Since their inception, the in-template version of these devices has suffered from surface quality control whereas the extruded version suffered from control of growth direction. Both of these issues now appear to be solvable. However, doping control remains a problem due to the inadvertent doping of the VLS catalyst. Purposeful doping to overcome this inadvertent doping may be the practical solution to this problem. These devices may have display applications and their simplicity and small size makes them very attractive for sensor applications.

1. Introduction

Silicon nanowires (SiNWs) have stimulated extensive research interest in recent years due to their potential as building blocks for nanoscale transistors, sensors and optoelectronic devices[1-3]. One of the most widely used techniques for producing SiNWs is the Au-catalyzed vapor-liquid-solid (VLS) process[1-6], which uses Au nanoparticles as the catalyst to grow SiNWs from precursor gases (e.g., $SiCl_4$ and SiH_4). There are generally two VLS approaches to growing SiNWs and positioning them[7-9]: (1) the grow-and-place approach and (2) the grow-in-place approach pioneered by our group. In our grow-in-place approach a template is used to form the SiNWs using the VLS technique. The SiNWs may be confined to the template (Fig. 1) or they may be allowed to extrude during growth (Fig. 2). Template formation for the grow-in-place approach has been discussed in detail in references 7-9. Generally speaking, the catalyst used in the VLS technique dopes the SiNWs. In the case of Au, the resulting doping is p-type. In this situation the wires can easily be made into accumulation transistor devices as shown schematically in Fig. 3. We have termed this transistor type an accumulation MOS field effect transistor (AMOSFET)[7, 8].

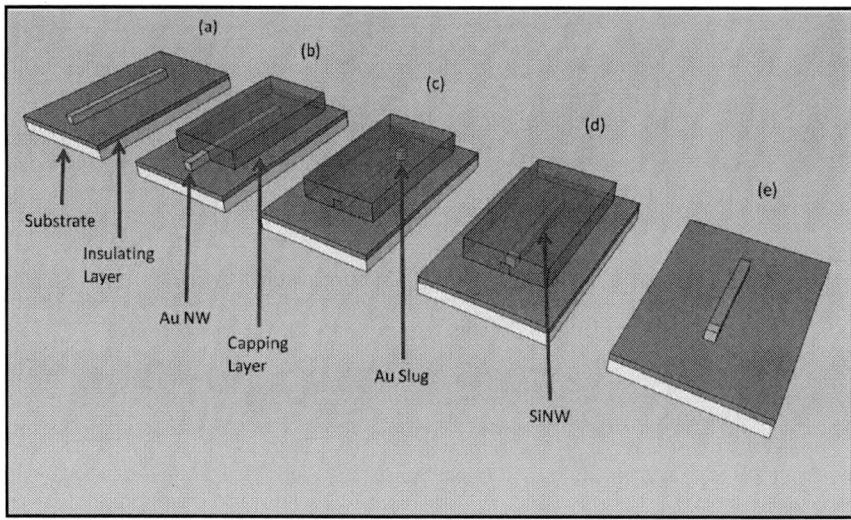

Figure 1. Grow-in-Place SiNWs confined to the growth template. Template formation and VLS catalyst (Au) positioning are shown.

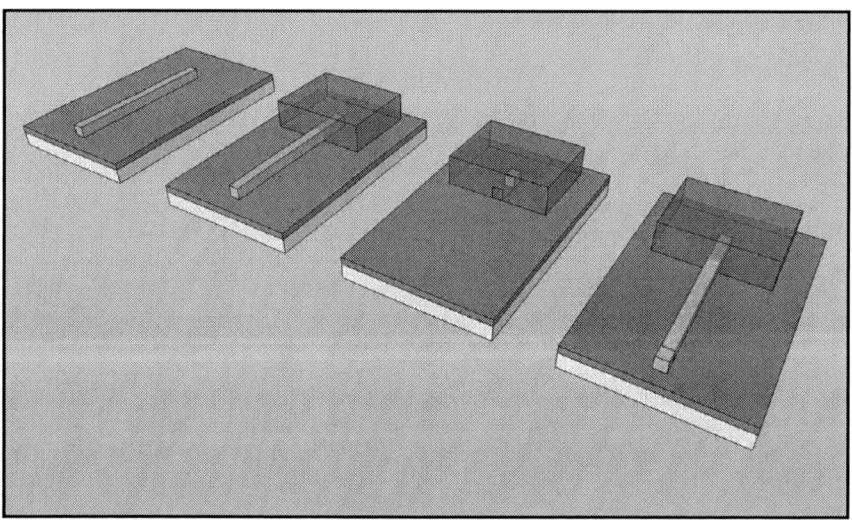

Figure 2. Grow-in-Place SiNWs allowed to extrude from the growth template. Template formation and VLS catalyst (Au) positioning are shown.

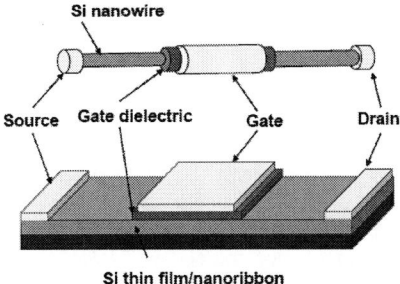

Figure 3. Schematic structures of AMOSFET devices for both SiNW and Si thin film. AMOSFETs requires only one doping type in which no junctions, such as p-n, n-i or p-i junctions, are present.

2. Advantages

The simplicity of the AMOSFET makes it an interesting device for electronics and for sensor applications. A table of its merits is found below.

Table I: Merits of AMOSFET

Merits
Could be a replacement for a pixel switch in AMLCDs, and would have a simple process with very low parasitic capacitance
No doping needed for the source-drain regions, simplifying the process
Gate dielectric thickness not critical
Low Off current can be achieved by full depletion of thin Si under gate
Gate-source and gate-drain capacitances can be very small
Low operating voltages and small sub-threshold swing

3. Issues

There are a number of problems that have been encountered with the grow-in-place approach which cause issues for AMOSFET utilization. One is the problem of surface roughness and polycrystalline wire formation for SiNWs grown within templates (i.e., not extruded). An example is seen in Fig. 4. It has been found that steps can be taken so that this problem no longer hampers non-extruded grow-in-place SiNWs.

While surface roughness and polycrystallinity are not problems with extruded grow-in-place SiNWs, there can be a growth control problem. This is seen in Fig 5 where the extruding SiNW is seen to have changed growth direction. It has been found that steps can be taken to rectify this problem too.

Inadvertent doping is an issue, in general, with VLS-grown SiNWs and consequently it hampers the grow-in-place approach also. Measurements of both unpassivated (as grown and cleaned) and passivated (as grown, cleaned, and oxidized) Au-catalyzed VLS "grown-in-place" SiNWs in our group has shown that the resistivities of these SiNWs are much lower than that of intrinsic silicon due to the use of the Au catalyst to seed the VLS growth. This inadvertent doping aspect of VLS growth does not present a problem for AMOSFETs since they are accumulation devices but doping concentration will have to be carefully controlled for applications.

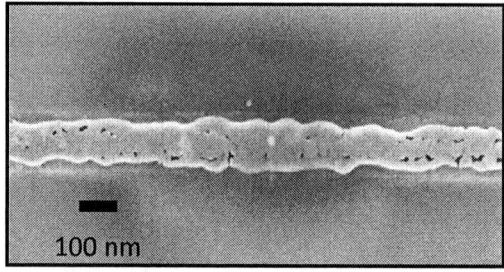

Figure 4. Surface roughness and porosity can occur for SiNWs within the growth template.

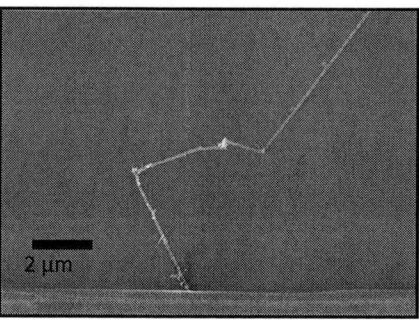

Figure 5. Growth direction changes can occur for SiNWs extruding from the growth template.

4. Conclusions

In 2007 our group produced SiNWs by the grow-in-place approach which was successfully used to fabricate devices with the proper gate-positioning layout to prevent ambipolar behavior and thin enough to deplete. Transistor action was demonstrated and the devices were termed accumulation metal oxide semiconductor accumulation field effect transistors (AMOSFETs). Subsequently, our group has also demonstrated planar AMOSFETs using thin film silicon. Issues such as surface quality had plagued the in-template version of these devices and control of growth direction had plagued the extruded version. Both of these issues now appear to be under control. However, doping control remains a problem due to the inadvertent doping of the VLS catalyst. Purposeful doping to overcome this inadvertent doping may be the practical solution to this problem.

REFERENCES

1. Y. Cui; Q. Q. Wei; H. K. Park; C. M. Lieber. *Science*, 293, 1289 (2001).
2. Y. Cui; Z. H. Zhong; D. L. Wang; W. U. Wang; C. M. Lieber. *Nano Letters*, 3 (2), 149 (2003).
3. M. C. McAlpine; R. S. Friedman; S. Jin; K. H. Lin; W. U. Wang; C. M. Lieber. *Nano Letters*, 3,(11), 1531 (2003).
4. O. Hayden; M. T. Bjork; H. Schmid; H. Riel; U. Drechsler; S. F. Karg; E. Lortscher; W. Riess. *Small*, 3 (2), 230 (2007).
5. J. B. Jackson; D. Kapoor; S. G. Jun; M. S. Miller. *Journal of Applied Physics*, 102 (5), 054310.1 (2007).
6. J. Y. Yu; S. W. Chung; J. R. Heath. *Journal of Physical Chemistry B*, 104 (50), 11864 (2000).
7. Y. Shan; S. Ashok; S. J. Fonash. *Applied Physics Letters*, 91 (9), 093518 (2007).
8. Y. Shan; S. J. Fonash. *Acs Nano*, 2 (3), 429 (2008).
9. Y. H. Shan; A. K. Kalkan; C. Y. Peng; S. J. Fonash. *Nano Letters*, 4 (11), 2085 (2004).

146

In-plane Silicon Nanowires for Field Effect Transistor Application

L. YU [a] and P. ROCA i CABARROCAS [a]

[a] Laboratoire de Physique des Interfaces et des Couches Minces (LPICM), Ecole Polytechnique/CNRS, 91128 Palaiseau, France

Silicon nanowires (SiNW) are important building blocks for a new generation of transistor and sensor applications. Integration and up-scaling of these functionalities rely on a critical ability to position and assemble these nanoscale 1D channels over large areas. We have proposed and demonstrated an in-plane solid-liquid-solid (IPSLS) growth mode of SiNWs, which enables unprecedented morphology control of the in-plane SiNWs and precise deployment of large-scale SiNW arrays. Notably, all the fabrication process can be accomplished in a CMOS compatible all-in-situ plasma deposition environment. We will present here the recent progress in this field and address particularly their related aspects for field effect transistor application.

Introduction

Low temperature plasma process have allowed the development of a-Si:H based TFTs for large area electronics. In particular a-Si:H based back-plane TFT arrays are the dominant technology for active matrix liquid crystal displays, despite their low mobility (~1 cm^2/V.s) and shift of the threshold voltage under electrical stress. Further, the crystallization of this material by various processes as well as the direct deposition of microcrystalline silicon films has been studied in order to overcome the limitations of a-Si:H TFTs, which do not make them suitable for integrated circuits (1). On the other hand, the high surface-to-volume ratio of silicon nanowires (SiNW) leads to a geometric enhancement in the capacitive coupling between the quasi-1D conductive channel and the gating electrode/environment, which improves the tunability and sensitivity of SiNW-based field effect transistors (FET) and sensors. This concept has laid the basis for developing a new generation of high performance FET and bio-sensors (2-5). SiNWs structure can be fabricated in either a top-down approach, by direct electro- or optical-lithography, or in a bottom-up approach through a self-assembly growth. The latter one boasts a low-cost and high-product approach for various SiNWs-based applications. For example, the well-known vapor-liquid-solid (VLS) (6-11) and oxide-assisted growth (OAG) (12-13) modes are widely used to fabricate SiNW structures with well-controlled diameter and morphology. In typical VLS or OAG modes, the growth of SiNWs carries on in a gas precursor environment and usually results in vertical arrays or bundles perpendicular to the substrate surface. Though novel vertical FET can be made directly on such vertical SiNWs forest, (14-15) reliable electrical connection and compatibility to mainstream CMOS are still challenging issues to be resolved for a realistic application. Actually, with the introduction of new nano-elements, the mature planar CMOS architecture is not going to be changed radically in a foreseeable future. To up-scale these promising SiNW functionalities, demonstrated mostly by plucking vertical VLS SiNWs and connected in a planar surface, the SiNWs need to be precisely positioned and addressable over large-scale. Particularly, it will be beneficial and convenient to accomplish all the fabrication steps, including the growth and positioning of SiNWs, in a CMOS compatible process/architecture. This is a particularly important for the concern of robustness and reliability in device integration.

The main difficulty in large-scale integration of SiNW-based devices is the incompatibility between the 3D growth mode of SiNWs, like VLS or OAG, and the 2D planar device architecture, which requires an extra re-arrangement or manipulation step to integrate vertical SiNWs into a 2D layout. From this point of view, a controllable in-plane growth of SiNWs mitigates the integration challenges in a planar architecture. Recently, we have proposed an in-plane solid-liquid-solid (IPSLS) mode to grow in-plane SiNWs, where indium drops transform hydrogenated amorphous silicon (a-Si:H) matrix into well-defined crystalline SiNWs, during an all-in-situ plasma processes in standard PECVD system (16-20). The low melting point of the indium catalyst drop, of 156 °C, enables a low temperature growth of the in-plane SiNWs, allowing the use of various low-cost and flexible substrates. We have demonstrated that this IPSLS growth mode provides unprecedented controllability over the morphology of the in-plane SiNWs structure (16-17). The growth path of in-plane SiNWs can be readily guided into desired position, automatically during the in-situ growth process. This enables precise "growth-in-place" control of the SiNWs channels for transistor applications (18-19), and is critical to implement a reliable device integration.

All-in-situ Fabrication and Morphology Control

All-in-situ Fabrication of In-plane SiNWs

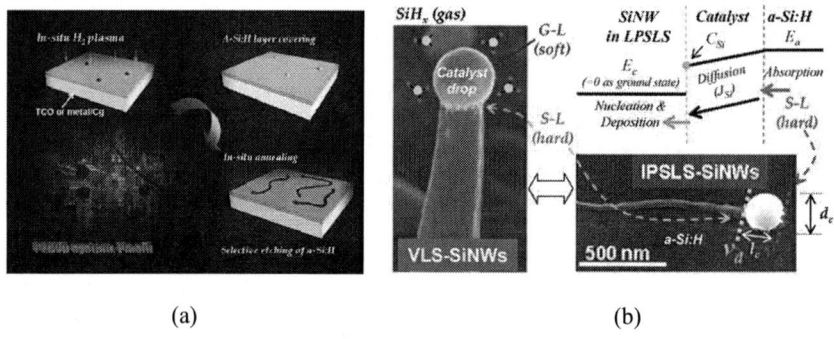

(a) (b)

Figure 1: (a) Illustration of the fabrication process of in-plane SiNWs, via IPSLS mode, in an all-in-situ plasma deposition process in a PECVD system; (b) Illustration for the different absorption and deposition interfaces involved in the vapor-liquid-solid (VLS) mode (on the left) and in the in-plane solid-liquid-solid (IPSLS) mode (on the right). The top-right inset depicts the relative Gibbs energy of Si atoms in a-Si:H matrix, catalyst drop and SiNWs, respectively.

Similar to a well-known VLS process, the growth of in-plane SiNWs is also catalyzed by nanoscale catalyst drops, except that the precursor used here is solid a-Si:H thin film instead of the gas phase silane molecules or radicals. The major fabrication process of the in-plane SiNWs are illustrated in Figure 1(a). First, we adopt an in-situ catalyst formation technique realized in a plasma enhanced chemical vapor deposition (PECVD) system. The precipitation of catalyst metal, indium (In) or indium-tin (In-Sn) alloy, and their agglomeration into droplets can be precisely controlled by tuning the H_2 plasma treatment of a thin layer of ITO at 150-250 °C. Typical H_2 flow rate, chamber pressure and RF power adopted in this catalyst formation step are 100 sccm, 600 mTorr, and 5 W respectively. Then, the samples are cooled down to 100 °C and followed by coating a thin layer of a-Si:H with a suitable thickness (10-50 nm), with corresponding plasma parameters of 10 sccm SiH_4 flow rate, 120 mTorr and 2 W. After the a-Si:H thin film deposition, the samples are annealed in vacuum ($< 10^{-6}$ mbar) or H_2 ambiance at 200 - 500 °C. During this process, the buried indium catalyst drops become molten again and start to absorb and transform the a-Si:H matrix into crystalline SiNWs, during their in-plane movement. Finally, the residual a-Si:H layer around the SiNWs can be selectively removed by a low temperature H_2 plasma etching at 100-130 °C, with H_2 flow rate, chamber pressure and RF power of 100 sccm, 20 W and 380 mTorr, respectively. All these fabrication steps can be done in an all-in-situ and one-pomp-down low-temperature process, as summarized in Table I below, and thus compatible to the standard CMOS or thin film transistor fabrication technology.

TABLE I. *Summary of Experimental Procedures for the all-in-situ fabrication process*

Fabrication Steps	Ambience	Temperature	In-situ Operation
1. Catalyst Formation	H_2 plasma	150-250 °C	√
2. a-Si:H Precursor Coating	SiH_4 plasma	<100 °C	√
3. Growth Annealing	Vacuum	200-500 °C	√
4. Remnant a-Si:H removal	H_2 plasma	<130 °C	√

Growth Dynamics and Morphology Control of In-plane SiNWs

The movement of the catalyst drops is driven by the difference in Gibbs energy of Si atoms in the amorphous (E_a) and crystalline (E_c) states, $\Delta E_{ac} = E_a - E_c$, as depicted by the top-right inset in Figure 1. The energy gain during this transformation process sustains the continuous crystallization of the a-Si:H matrix and the growth of SiNWs behind. In contrast to a VLS process, the front solid-liquid (S-L) interface in the IPSLS process is a hard interface, compared to the soft gas-liquid (G-L) interface seen in the VLS process as indicated in Figure 1. During the growth, the movement of the front absorption interface (indicated by red dashed line in Figure 1) is effectively coupled to the rear deposition interface (indicated by the green dashed line). In general, these two interfaces have to move in pace, otherwise the liquid catalyst drops in between will be deformed. As a consequence of this unique moving-rate-coupling in IPSLS process, which is totally absent in the VLS process, the morphology of the SiNWs becomes readily controllable during their in-situ growth.

In formulating the interface interplay during an IPSLS process, we define the moving rates of the front (absorption) and the rear (deposition) interfaces in a running catalyst drop as v_d and v_a, respectively. According to mass conservation, the incoming absorption Si flux should be balanced by the out-going deposition flux,

$$v_a \cdot h_a \cdot d_c \cdot \alpha = v_d \cdot d_w^2 \Rightarrow \eta \equiv v_a / v_d = f \cdot d_c /(h_a \cdot \alpha),$$ [1]

where $f \equiv (d_w / d_c)^2$ and $\alpha \equiv \rho_{a-Si} / \rho_{c-Si}$ are geometric and density ratio constants, and h_a and d_w stand for the thickness of a-Si:H layer and the diameter of the produced SiNW, respectively. As we can see, given a specific h_a, there is only one catalyst drop diameter d_c (or $f^2 \cdot d_w$) that will satisfy a balance condition of $\eta = v_a / v_d \sim 1$, that is, we can define a critical catalyst size of $d_{cr} \equiv h_a \cdot \alpha / f$. For a spherical large (small) catalyst drop with initial diameter $d_c > d_{cr}$ ($d_c < d_{cr}$), the catalyst drop will be stretched (squeezed) since $\eta = v_a / v_d = d_c / d_{cr} > 1$ ($\eta = v_a / v_d = d_c / d_{cr} < 1$), and thus produce straight (bending) SiNWs, as seen in Figure 2.

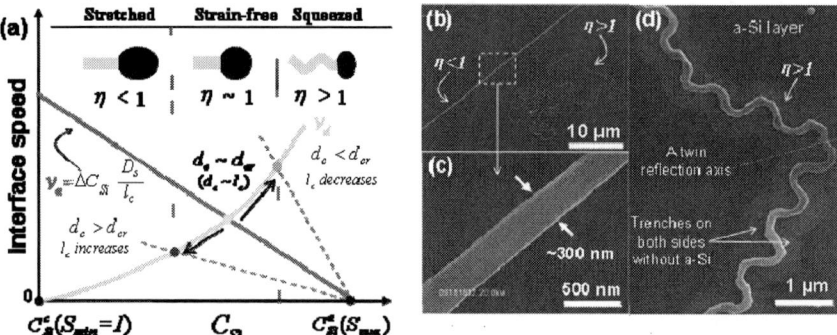

Figure 2. (a) Speeds of the front absorption (v_a) and the rear deposition (v_d) interfaces, as a function of the dissolved Si concentration at the SiNW/catalyst interface. The top insets indicate the three possible strain states and corresponding deformations of the catalysts, as well as the typical SiNWs produced under the different growth balance conditions. (b) SEM images of a thick SiNW (~300nm) grown under a condition $\eta < 1$ and of a thin SiNWs (~150 nm) grown under $\eta > 1$, with the enlarged view of the straight SiNW presented in (c). (d) shows a zigzag SiNW with regular growth orientations switching obtained under a condition of $\eta > 1$.

Furthermore, the nucleation and absorption processes are also related to the dissolved Si atom concentration C_{Si} and the supersaturation $S \equiv C_{Si} / C_{Si}^{eq}$ established in the catalyst drop (in vicinity to the deposition interface), where C_{Si}^{eq} is the equilibrium Si concentration at the c-Si/In interface. v_d increases monotonically with the increasing of S (or C_{Si}), in a way exactly the same to the already known nucleation/deposition process at the catalyst/SiNWs interface in a VLS process (21) while, the absorption rate v_a, at the front a-Si:H/catalyst interface, is basically a diffusion-flux-limited (under the current experiment conditions) (16-17). In a catalyst drop of length l_c, we have

$$v_a \sim (C_{Si}^a - C_{si}) \cdot D_s / l_c = (C_{Si}^a - S \cdot C_{Si}^{eq}) \cdot D_s / l_c, \qquad [2]$$

with $C_{Si}^a = C_{Si}^{eq} \cdot e^{\Delta E_{ac}/kT}$ being the Si concentration at the a-Si:H/catalyst interface, and D_s the diffusion coefficient of Si atom in the catalyst liquid. As shown in Figure 2(a), C_{Si} is actually determined by the intersect of the two curves (at $v_d = v_a$), and the growth rate of the in-plane SiNWs will be faster for thin wires than that of the thicker ones. This is indeed supported by our in-situ SEM observation. More detailed explanation and discussion of this point are available in our previous works (17).

Based on this unique growth balance condition during the in-plane growth of SiNWs via IPSLS mode, we are able to design the morphology of the in-plane SiNWs without the use of any constraint template or channels. This can be seen, for example, by the different morphologies of the SiNWs with different diameters (and thus the values of η). We note that, this ability is unachievable in VLS and OAG growth modes in a gas phase precursor environment. The morphology tunability of the in-plane SiNWs of different diameters is assured by the fact that the ratio of $\eta \sim d_c / h_a$ is the only controlling factor, while both the a-Si:H thin film deposition (for h_a) and the catalyst formation (for d_c) can be precisely and readily controlled in a plasma deposition process. It is also worthy to note that a SiNW channel with regular zigzag shape as seen in Figure 2(d), due to the coupling of preferential crystalline orientation in c-Si to the growth dynamics, could bear the potential to develop a flexible and yet stable and high performance (high crystallinity of the SiNWs) thin film transistor application.

Large-scale Self-position and Bottom-gate FET Performance

Guided Growth of In-plane SiNWs

The IPSLS growth mode enables also an all-in-situ self-positioning of SiNWs in large-area arrays. This unique ability derives from a guided growth mechanism (19), which relies only on a simple step-edge defined on a planar surface. For example, as seen in the SEM images in Figure 3 (a), parallel in-plane SiNWs can be grown and positioned automatically along the guiding step-edges, with precisely controlled size and number of SiNWs. A schematic picture of the overall structure for the production of bottom gate SiNW TFTs is illustrated by Figure 3 (b). The fabrication steps are:

i)   A thin layer of ITO (~10 nm thick) was deposit and patterned into stripe rows of widths 1~5 μm, on top of an n⁺-Si wafer coated with a 300 nm thick SiO₂ layer, serving as a gate dielectric.

ii)   ii) An a-SiN$_x$ layer (~180 nm thick) was deposited and window channels of width of 20 μm were opened (being normal to the ITO rows) to serve as the guiding edges.

iii)   the samples were loaded into a PECVD system, and the typical growth conditions, as described above, were applied to produce the in-plane SiNWs.

Here, since the catalyst drops will be formed only on top of the exposed ITO pads at the crossing point, the initial growth position of the in-plane SiNWs was thus positioned automatically close to the end of the guiding edges. When a catalyst drop runs into the step-edge, a new absorption front, as indicated by the green line in Figure 3 (c), is formed on the vertical step sidewall. This new absorption front will exert an effective attraction force on the moving catalyst drops, and constrain the catalyst drop to move exactly along the step edges, producing one SiNW along each guiding edge.

iv) The self-aligned in-plane SiNWs were then provided with two Al electrodes at the two ends, serving as source and drain contact, while the common bottom gate electrode is provided by the n^+-Si substrate.

The transfer characteristic of a SiNW FET, with a length of 200 μm and a diameter of 170 nm is presented in Figure 3 (d). Though no intentional doping source is introduced during the fabrication process, the transfer behavior indicates a p-type doped SiNW channel. This is due to the incorporation of In atoms during growth, which is known to introduce p-type doping in c-Si. (22) This prototype SiNW-FET shows an on/off ratio $>10^3$ and a sub-threshold swing of 690 mV/dec. The hole mobility extracted from the bottom-gate configuration, as illustrated in the inset of Figure 3(d), is around 228 cm²/V.s. These results indicate a new opportunity to deploy high performance SiNWs applications in a CMOS compatible and yet cost-efficient way.

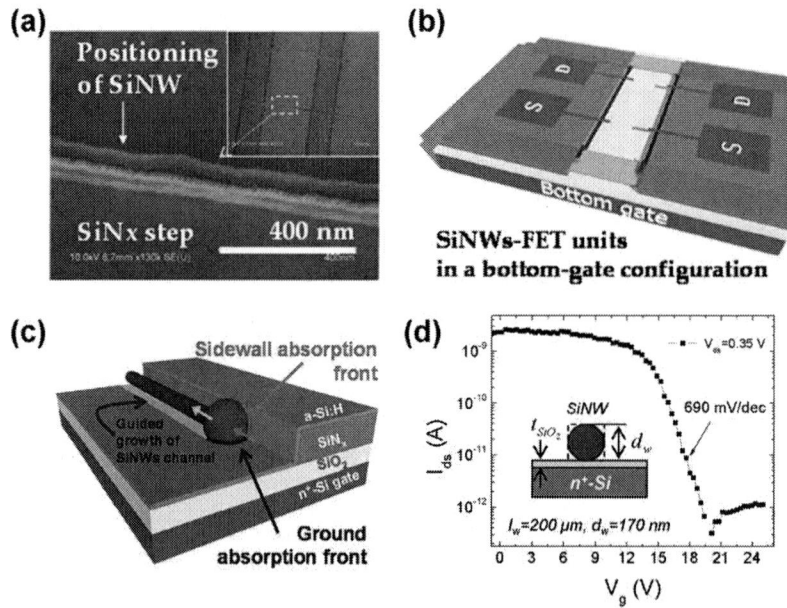

Figure 3: (a) SEM image of the guided growth of in-plane SiNWs along the pre-defined guiding edges. (b) Schematic illustration of the SiNW-FET structure in a bottom-gate configuration, based on the guided SiNWs. (c) Illustrates the guided growth mechanism

of the in-plane SiNWs by a simple guiding edge. (d) Transfer properties of the SiNW FET prototype, with the geometrical parameters of the measured structure depicted in the inset.

Summary

We have reviewed the unique features of SiNWs grown in a newly proposed in-plane solid-liquid-solid (IPSLS) mode, in terms of morphology and precise growth position/path control. These features provide the important basis and have been successfully applied to deploy self-positioned in-plane SiNWs for field effect transistor applications. Prototype devices fabricated in a simple bottom-gate configuration have been successfully realized and demonstrated an on/off ratio $>10^3$, carrier mobility of 228 cm^2/Vs and sub-threshold slope of 690 meV/dec. These results open up new opportunities and lay an important basis for direct integration of various SiNWs-based applications.

References

1. M. Oudwan, O. Moustapha, A. Abramov, D. Daineka, Y. Bonnassieux, and P. Roca i Cabarrocas. Phys. Stat. Solidi A **207**, 1245 (2009).
2. Y. Cui, Q. Q. Wei, H. K. Park, and C. M. Lieber, *Science* **293**, 1289 (2001).
3. J. F. Hsu, B. R. Huang, C. S. Huang, and H. L. Chen, *Japanese Journal of Applied Physics Part 1-Regular Papers Brief Communications & Review Papers* **44**, 2626 (2005).
4. Y. Y. S. N. B. W. S. T. L. M.-W. Shao, *Advanced Functional Materials* **15**, 1478 (2005).
5. S. Roy, and Z. Gao, *Nano Today* **4**, 318 (2009).
6. R. S. Wagner, and W. C. Ellis, *Appl. Phys. Lett.* **4**, 89 (1964).
7. Y. Cui, L. J. Lauhon, M. S. Gudiksen, J. Wang, and C. M. Lieber, *Appl. Phys. Lett.* **78**, 2214 (2001).
8. Y. Wu, *Nano Lett.* **4**, 433 (2004).
9. A. Colli, A. Fasoli, P. Beecher, P. Servati, S. Pisana, Y. Fu, A. J. Flewitt, W. I. Milne, J. Robertson, C. Ducati, S. De Franceschi, S. Hofmann, and A. C. Ferrari, *J. Appl. Phys.* **102**, 034302 (2007).
10. J. F. Dayen, A. Rumyantseva, C. Ciornei, T. L. Wade, J. E. Wegrowe, D. Pribat, and C. S. Cojocaru, *Appl. Phys. Lett.* **90**, 173110 (2007).
11. Y. Shan, and S. J. Fonash, *ACS Nano* **2**, 429 (2008).
12. R.-Q. Zhang, Y. Lifshitz, and S. T. Lee, *Adv. Mater.* **15**, 635 (2003).
13. N. Wang, Y. H. Tang, Y. F. Zhang, C. S. Lee, and S. T. Lee, *Phys. Rev. B* **58**, R16024 (1998).
14. S. Nam, X. Jiang, Q. Xiong, D. Ham, and C. M. Lieber, *Proceedings of the National Academy of Sciences* **106**, 21035 (2009).
15. V. Schmidt, H. Riel, S. Senz, S. Karg, W. Riess, and U. Gösele, *Small* **2**, 85 (2006).
16. L. Yu, P.-J. Alet, G. Picardi, and P. Roca i Cabarrocas, *Phys. Rev. Lett.* **102**, 125501 (2009).
17. L. Yu, and P. Roca i Cabarrocas, *Phys. Rev. B* **81**, 085323 (2010).

18. L. Yu, O. Moustapha, M. Oudwan, and P. Roca i Cabarrocas, *Mater. Res. Soc. Symp. Proc.* **1178E**, AA07 (2009).

19. L. Yu, M. Oudwan, O. Moustapha, F. Franck, and P. Roca i Cabarrocas, *Appl. Phys. Lett.* **95**, 113106 (2009).

20. L. Yu, and P. Roca i Cabarrocas, *Phys. Rev. B* **80**, 085313 (2009).

21. V. Schmidt, J. V. Wittemann, and U. Gösele, *Chemical Reviews* **110**, 361 (2010).

22. S. M. Sze., *Physics of Semiconductor Devices* (Wiley, New York, 1981), Vol. 1.

CHAPTER 4

CHALLENGES IN MEMORIES

156

Nonvolatile Memories for Nano and Giga Electronics

Yue Kuo

Thin Film Nano & Microelectronics Research Laboratory, Texas A&M University, College Station, TX 77843-3122, U.S.A.

Memory devices are critical elements in ULSIC. It is also desirable to include memory functions in a-Si:H TFTs for new and novel applications. In this paper, the author reviews and discusses recent developments on two types of nonvolatile memories in his laboratory, i.e., the nanocrystals embedded high-k MOS capacitor and the floating-gate a-Si:H TFT. For the former, the structure, device performance, and reliability are examined. For the latter, charge trapping and detrapping mechanisms and memory functions are investigated. Challenges of these devices toward nano and giga electronics are analyzed.

Nonvolatile Memories for Nano and Giga Electronics

Development trends of ULSICs and TFTs are opposite. The former is toward squeezing a tremendous number, e.g., a trillion or more, of nano-size devices into a very small die area. The latter is to expand the panel size, e.g., over 100 inches diagonally, on the extremely large substrate, e.g., > 3 m by 3 m. Although it is desirable to shrink the transistor size in the direct view display for better resolution and power consumption, there is no need of fabricating nanometer TFTs as long as the picture quality is satisfactory. In spite of the above difference, basic device operation principles of the MOSFET and the TFT are the same, i.e., both are field effect transistors.

When the MOSFET size is reduced to the nano size, the SiO_2 gate dielectric needs to be replaced with a high-k material. A high-k dielectric film with the same equivalent oxide thickness (EOT) as SiO_2 can reduce the leakage current and improve reliability. For future nano size memory devices, high-k dielectrics have also been used to satisfy stringent requirements on size, power, speed, reliability, etc. The publication numbers related to high-k memories increases monotonically in recent years (1). High-k films have been used in ONO, resistive switching, phase change, DRAM, SRAM, or nanocrystals charge trapping memory devices for their unique material or structure properties (2-7). The author's group has done extensive research on the nanocrystals embedded high-k dielectric nonvolatile memories that show promising results (8,9,10). The combination of nanocrystalline dots and the high-k dielectric could enhance the charge storage capacity, lower the operation power, and improve the data writing speed (11).

Traditionally, memory functions are rarely considered in the TFT field due to the focused application to the pixel-level liquid crystal driving. Recently, it has been reported the organic light emitting diode (OLED) can be driven with a-Si:H TFTs and capacitors

(12). The a-Si:H TFTs have also been used in chemical and biological sensing (13). In many of these applications, new functions can be introduced into the product if the a-Si:H TFT-based memory device is available. Recently, the author's group demonstrated that the floating-gate a-Si:H TFT and capacitor can be fabricated into nonvolatile memories (14-18). Since the a-Si:H TFT can be mass-produced on various types of rigid and flexible substrates at a low cost, this kind of memory device is easily integrated into displays, sensors, imagers, etc.

Principles of Operation and Device Structures

The nanocrystals embedded high-k MOSFET and the a-Si:H embedded floating-gate TFT memory devices are based on the same operation principle. Figure 1 shows the band diagram of a floating-gate MOS capacitor at the equilibrium state, i.e., non-stressed condition. As long as the dielectric layers attract negligible amount of charges, in principle, any materials with a large conduction or valence band offset with the tunnel (or channel-contact) dielectric can be used as the charge trapping layer. It can be in either discrete or continuous form.

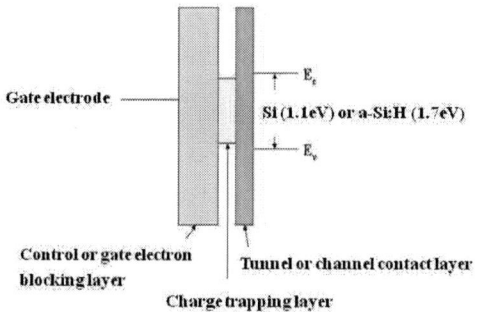

Fig. 1. Band diagram of a floating-gate MOS capacitor.

For the floating-gate a-Si:H TFT or capacitor, the complete semiconductor channel layer and gate dielectric stack could be deposited by plasma enhanced chemical vapor deposition (PECVD) on a glass or flexible substrate in one pump down, which prevents environment contamination. Typically, the channel layer is composed of an a-Si:H thin film; the channel-contact and control dielectrics are made of SiN_x; the embedded layer is an ultra-thin a-Si:H film. The Fig. 1 diagram well represents the actual device band structure because the low temperature, e.g., $\leq 350°C$, PECVD process warrants the atomic sharp interface between a-Si:H and SiN_x.

The nanocrystals embedded high-k stack can be prepared with various thin film deposition methods. In our study, the complete dielectric stack was deposited by sputtering in one pump down without breaking the vacuum. This simple process is easily transferred to large-area wafers for mass production. The as-deposited embedded layer is usually a very thin continuous film, which is transformed to crystalline nano dots during the subsequent post deposition annealing (PDA) step. The Fig. 1 diagram needs to be modified to take into consideration of the two interfaces formed between the tunnel dielectric and the Si substrate as well as between the control dielectric and the gate electrode. Their properties and thicknesses are related to the high temperature, e.g., 800-1,000°C, PDA and the post metal annealing, e.g., 400°C, conditions (19). The PDA process is critical to properties of the embedded nanocrystals. The final annealing step is necessary to repair defects generated during the deposition of the gate electrode and the back ohmic metal. An interface layer between the nanocrystal and the surrounding high-k can also be formed (20).

Figure 2 shows the band diagrams of a nanocrystalline ruthenium oxide (nc-RuO) embedded Zr-doped HfO$_2$ (ZrHfO) capacitor under the (a) +V_g bias, i.e., electron-injection, and (b) -V_g bias, i.e., hole-injection, conditions, separately. It contains a HfSiO$_x$ interface layer between the tunnel ZrHfO and Si and an Al$_2$O$_3$ interface layer between the control ZrHfO and the aluminum (Al) gate (19). The conduction- or valence-band offset between the tunnel ZrHfO and the Si determines the energy required to "write" the data. Similarly, the band offset between nc-RuO and ZrHfO affects the energy required to "erase" the data. The charge retention property depends on the band offset, the quality and thickness of the tunnel ZrHfO film, the operation temperature, etc. (21).

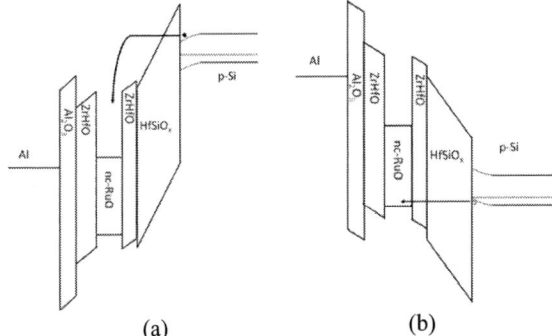

(a)   (b)

Fig. 2. Band diagrams of an nc-RuO embedded ZrHfO capacitor on a p-type Si wafer at (a) electron- and (b) hole-injection conditions.

Nanocrystals Embedded High-k Nonvolatile Memory Device Characteristics

The charge trapping density (Q_{ot}) of a MOS capacitor can be estimated from the flat band voltage shift (ΔV_{FB}) of the C-V hysteresis curve, i.e., V_g swept from one polarity to another polarity and then back, or the C-V curve shift before and after a V_g stress. The following equation is commonly used for the Q_{ot} calculation (10):

$$Q_{ot} = \frac{C_{FB} \times \Delta V_{FB}}{q} \tag{1}$$

where q is the electron charge and C_{FB} is the flatband capacitance. Q_{ot} is influenced by the operation parameters, such as the magnitude and direction of the sweep or stress V_g, time, and temperature, as well as the physical properties of the capacitor, e.g., the nano dot material and density (8,22-25). For example. Figure 3 shows changes of the charge trapping capacities of the single- and dual-layer nc-ITO embedded ZrHfO capacitors on the p-type Si wafer with respect to the magnitude of the stress V_g and time (20). In this case, the negative V_g is applied because holes are preferably trapped to the nc-ITO site (10). Figure 4 shows that the dual-layer sample has a nano dot density much higher than that of the single-layer sample, i.e., 1.13×10^{13} vs. 2.83×10^{12} cm^{-2}, which is consistent with the former's larger charge trapping capacity (23).

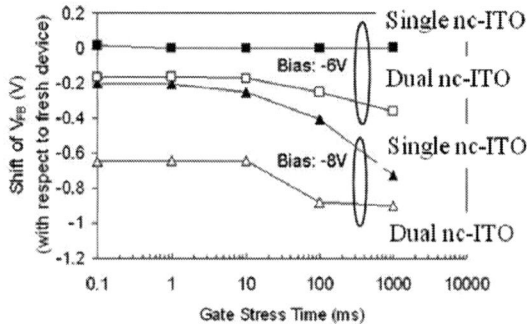

Fig. 3. Time and stress V_g magnitude effects on the V_{FB} shift with respect to the single- and dual-layer nc-ITO embedded ZrHfO capacitors (19).

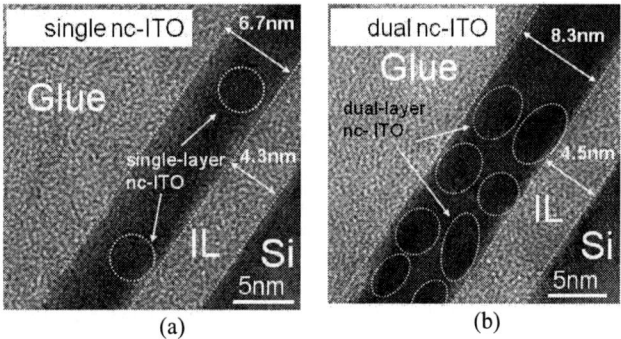

Fig. 4. Cross-sectional TEM of (a) single- and (b) dual-layer nc-ITO embedded ZrHf (23)

The type of the embedded nanocrystalline material determines the device's preference for electron or hole trapping (2,12,39,40). In addition, charges can be strongly or weakly trapped within the bulk nanocrystal or at the nanocrystal/high-k interface. These trapping states can be differentiated using the frequency-dependent C-V or G-V measurement (20,26). The loosely trapped charges are released at a slower rate than those trapped in the bulk high-k film that has the polarizable bond structure (27). When the bulk high-k film is broken down, the trapped charges could be retained for a period of time judged from the relaxation current's polarity change (28). However, the detrapping process is accelerated with the increase of the temperature. The high temperature also deteriorates the capacitor's charge trapping capability through mechanisms, such as increasing the tunnel dielectric layer's leakage current or the interface density of states (D_{it}) (29).

The reliability of the nanocrystals embedded ZrHfO capacitor was studied using the constant voltage stress (CVS) and stress induced leakage current (SILC) methods at different temperatures (23,27,30,31). Basically, charges are transferred through the high-k stack through the field emission mechanism at the low electric field but the Poole–Frenkel emission at the high electric field (30). The long stress time weakens the high-k film, which causes the early start of the Poole-Frenkel mechanism.

In summary, the nanocrystals embedded high-k structure shows excellent nonvolatile memory functions. On the average, each 3-5 nm size nano dot traps about one charge. For the future nano size transistor, it is possible to form one nanodot in the channel region by controlling the sample preparation process. Then, the nonvolatile memory function can be dependent on the charge and discharge of single electron or hole.

Floating-Gate a-Si:H Capacitor and TFT Memory Device Characteristics

The floating-gate a-Si:H MIS capacitor has been fabricated and studied (32). Compared with the nanocrystals embedded high-k MOS capacitor, all films are thicker and the required operating voltages are higher. Figure 5 shows (a) the cross-sectional view and (b) the C-V hysteresis curve of an a-Si:H floating-gate capacitor (32). A ΔV_{FB} of 2.1 V is obtained in the C-V hysteresis curve, which corresponds to a charge trapping density 2 orders of magnitude higher than that of the control sample, i.e., without the embedded a-Si:H layer in the SiN_x gate dielectric. However, the floating-gate capacitor shows more serious frequency-dependent capacitance change than the control sample in the depletion region due to the existence of the additional embedded a-Si:H/channel contact interface (15).

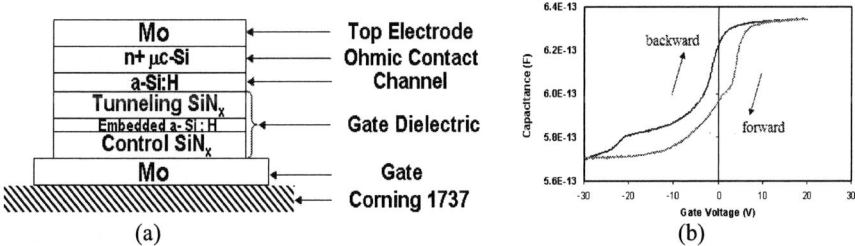

Fig. 5. (a) Cross-sectional view (c) C-V hysteresis of an a-Si:H floating-gate capacitor (32).

For the floating-gate a-Si:H TFT, the charge storage density (Q) can be estimated from the threshold voltage shift (ΔV_t) of the hysteresis of the transfer characteristics curves or the difference between the stressed and the fresh, i.e., unstressed, transfer characteristics curves. The following equation is often used for the calculation (33):

$$Q_{ot} = \frac{\Delta V_t \varepsilon}{d} \tag{2}$$

where ε and d are the dielectric constant and the thickness of the channel-contact SiN_x layer, respectively. Figure 6 shows examples of the transfer hysteresis curves of (a) a control TFT, i.e., without the embedded a-Si:H layer, and (b) an a-Si:H floating-gate TFT (34). The control sample corresponds to a very small Q while the floating-gate TFT corresponds to a large Q, i.e., 8.9×10^{11} vs. 1.7×10^{13} C/cm^2.

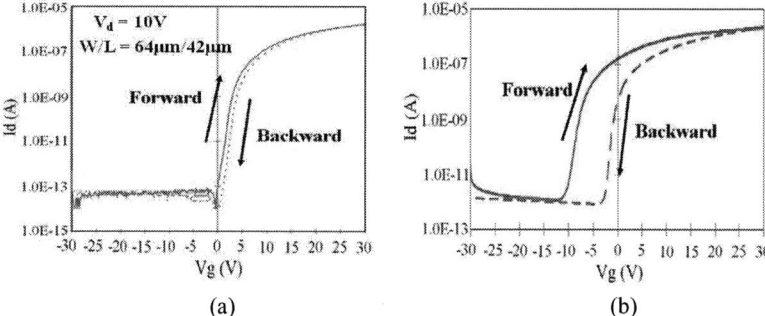

(a)   (b)

Fig. 6. Hysteresis of transfer characteristics curves of (a) a control TFT and (a) a floating -gate TFT with a 150 nm channel-contact SiN_x layer, ~13 nm embedded a-Si:H layer, and 150 nm control SiN_x layer. V_g: -30V to 30V to -30V. V_d = 10V. W/L = 64/42 μm (34).

The charge trapping capacity is dependent on the operation condition, e.g., magnitude of stress V_g, stress time, and temperature and the physical structure, e.g., the channel length of the TFT or the location of the embedded a-Si:H layer (15,34). For example, the Q value increases with the increase of the stress V_g because of the increase of the supply of the high energy charges (16). The charge storage density follows the power law

$$Q = Q_0 t^\gamma \tag{3}$$

where Q_0 is the initial charge storage density before stress, t is the stress time, and γ is a fitting coefficient. This relationship is consistent with dangling bond concentration that also increases with the stress time following the power law (35). When biased with the same V_g for the same period of time, the short-channel floating-gate TFT has a larger charge storage density than the long-channel floating-gate TFT (35). This may be related to the channel resistance or the dimension of the gate-to-source overlap area.

The location of the embedded a-Si:H in the gate dielectric layer is critical to the charge trapping and detrapping mechanisms (36). The channel-contact SiN_x layer thickness and the magnitude of the sweep V_g affect the locations and shapes of the forward and backward transfer characteristics curves, which then determines the memory capacity (36,37,38). For example, Figure 7 shows that by keeping the total gate SiN_x thickness fixed at 300nm, the transfer characteristics curve, i.e., with V_g swept from -30V to 30V, shifts toward the negative V_g direction with the thinning of the channel-contact SiN$_x$ layer, i.e., from 100 nm to 200nm (36). At the same -V_g, the thin channel-contact SiN$_x$ layer favors the formation of the hole-rich inversion layer and their subsequent injection into the gate dielectric stack. Therefore, the curve shifts to the negative V_g direction. The bump in the curve becomes more obvious with the decrease of the channel-contact SiN$_x$ layer. For the floating-gate TFT with a thin channel-contact SiN$_x$, a large portion of the trapped holes are loosely retained at the embedded a-Si:H site. They are quickly released when the V_g is reduced to near the V_t. This phenomenon is less obvious when less holes are injected to the gate stack, e.g., with a small starting -V_g (38).

The starting V_g of the backward transfer characteristics curve, i.e., swept from positive to negative, influences the electron trapping states. The low V_g trapped electrons are easier to detrap than the large V_g trapped electrons, as shown in Figure 8 (36).

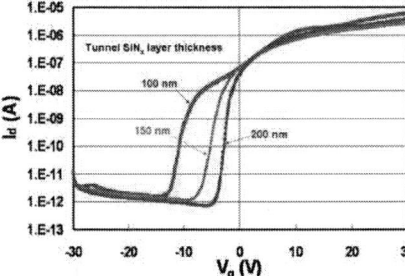

Fig. 7. Transfer characteristics of floating-gate TFTs with channel-contact SiN$_x$ layer thickness 100, 150, and 200 nm, separately. Starting V_g -30V. W/L = 59/21μ (36).

Fig. 8. Backward curves of the TFT with 100 nm channel-contact SiN$_x$ layer. Forward curves -30V to 30V, -20V to 20V, and -10V to 10V, separately. W/L = 59/21μ (36).

The combination of the location of the embedded a-Si:H layer and the sweep V_g range determines the direction and shape of the transfer hysteresis curve. For example, the forward to backward hysteresis can be counterclockwise, clockwise, or even cross, as shown in Figure 9(a), (b), and (c) (37). Therefore, in order to achieve a large memory window, these parameters need to be optimized.

The trapped charges on the a-Si:H floating-gate TFT can be released with various methods, as shown in Figure 10 (16). Majority of the trapped electrons could be detrapped with a thermal annealing, negative V_g stress, or light exposure method. They either provide extra energy to the trapped charges to overcome the band offset or to induce the current leakage path in the channel-contact SiN$_x$ layer so that charges can be released to the channel region (39-43).

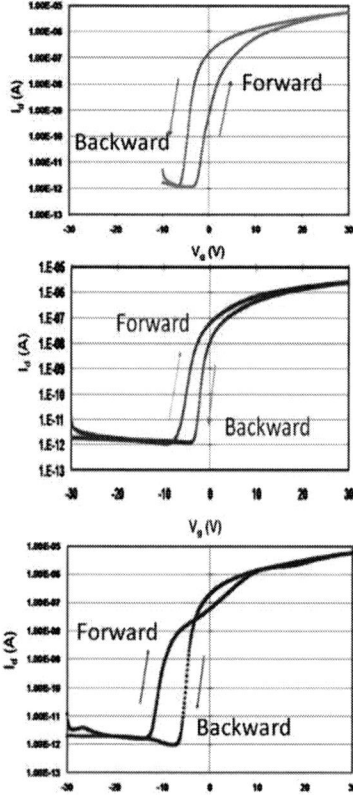

Fig. 9. Floating-gate TFTs with channel-contact SiN$_x$ layers (a) 100 nm thick and V_g -10V to 30V to -10 V, (b) 150 nm thick and V_g -30V to 30V to -30 V, and (c) 100 nm thick and V_g -30V to 30V to -30 V (37).

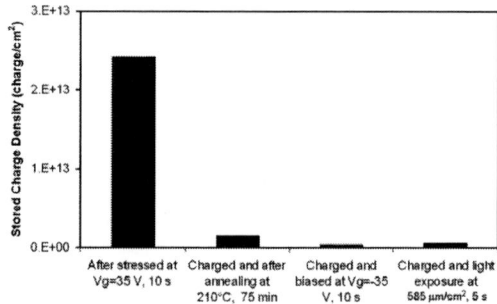

Fig. 10. Charge densities of the floating-gate a-Si:H TFT after stress and various subsequent treatments. W/L =93/23 µm (16).

In summary, the floating-gate a-Si:H TFT is an effective nonvolatile memory device that can be fabricated on various large-size, low-temperature rigid or flexible substrates. It can have a charge storage density higher than that of a conventional poly-Si floating-gate MOSFET. The stored charges can be released using electrical, thermal, and photonic methods. However, the gate dielectric structure and the V_g stress condition have to be optimized to store the maximum number of charges.

Conclusions

Two types of nonvolatile memories have been discussed. The nanocrystals embedded high-k gate dielectric structure is suitable for the future generation IC product. When the device is shrunk to the nanometer size, this kind of device may be operated based on the trapping and detrapping of a single electron or hole. The floating-gate a-Si:H TFT is applicable to products that are not restricted to substrate size or materials. This kind of device is easy to fabricate using the current a-Si:H TFT process. The new memory device can be included in the existing flat panel display to add new functions or added to novel sensors, solar cells, etc.

Acknowledgments

The author would like to thank his current and past graduate students for fabrication and characterization of devices discussed in this paper. He also acknowledges NSF CMMI-0926379 and CMMI-0654172 project supports as well as DHS grant numbers 2009-DN-077-ARI018-03 and 2008-DN-077-ARI018-04.

References

1. Y. Kuo, *ECS Trans.*, **33**(3), 425 (2010).
2. T.-H. Hsu, H.-C. You, F.-H. Ko, and T.-F. Lei, *J. Electrochem. Soc.,* **153**(11) G934 (2006).
3. H.-T. Lue, S.-C. Lai, T.-H. Hsu, P.-Y. Du, S.-Y. Wang, K.-Y. Hsieh, R. Liu, and C.-Y. Lu, *47th IEEE IRPS Symp. Proc.*, 874 (2009).
4. P. Blomme, B. Govoreanu, M. Rosmeulen, A. Akheyar, L. Haspeslagh, J. De Vos, M. Lorenzini, J. Van Houdt, and K. De Meyer, *ECS Trans.*, **1**(5) 75 (2006).
5. Y.-R.Tsai, K.-C. Liao, S. Maikap, *ULSI 2009*, 229 (2009).
6. R. F. Tsui, J. R. Shih, K. Liu, Y. S. Tsai, H. W. Chin, K. Wu, *VLSI Tech.*, 1-2 (2006).
7. Y. Wang, U. Bhattacharya, F. Hamzaoglu, P. Kolar, Y. Ng, L. Wei, Y. Zhang, K. Zhang, M. Bohr, *ISSCC* 2009, **456-457**, 457a (2009).
8. C.-H. Lin and Y. Kuo, *Electrochem. Solid State Lett.,* **13**(3), H83 (2010).
9. J. Lu, Y. Kuo, J. Yan, and C.-H. Lin, *Jpn. J. Appl. Phys.*, **45**(34), L901 (2006).
10. A. Birge and Y. Kuo, *J. Electrochem. Soc*, **154 (10)**, H887 (2007).
11. J. J. Lee, X. Wang, W. Bai, N. Lu and D. L. Kwong, in *IEEE Trans. Electron Dev.*, **50**(10), 2067 (2003).
12. J.-C. Goh, J. Jang, K.-S. Cho, and C.-K. Kim, *IEEE Electron Device Lett.*, 24(9), 583 (2003).
13. Y, Kuo, Chapt. 11, *Thin Film Transistors, Materials and Processes, Volume 1:*

Amorphous Silicon ThinFilm Transistors, New York, Kluwer, *pp.* 485 (2004).

14. Y. Kuo and H. Nominanda, *Appl. Phys. Lett.* **89**, 1 (2006).
15. Y. Kuo and H. Nominanda, *MRS Proc.,* **989**, A10-03 (2007).
16. Y. Kuo and H. Nominanda, *J. Korean Phys. Soc.* **54**(1), 409 (2009).
17. Y. Kuo and H. Nominanda, *MRS Proc.,* **1066**, A08-02 (2008).
18. H. Nominanda and Y. Kuo, *Electrochem. Solid State Lett.,* **10**(8), H232 (2007).
19. C.-H. Lin and Y. Kuo, to be published.
20. C.-H. Lin and Y. Kuo, *ECS Trans.,* **35**(2) 249 (2011).
21. C.-H. Yang, Y. Kuo, C.-H. Lin, and W. Kuo, *Electrochem. Solid State Lett.,* **14**(1) H50-H52 (2011).
22. C.-H. Lin and Y. Kuo, *ECS Trans.,* **13**(1), 465 (2008).
23. C.-H. Lin, C.-H. Yang, and Y. Kuo, *ECS Trans.,* **28**(1), 269 (2010).
24. C.-H. Lin and Y. Kuo, *ECS Trans.,* **19**(8) 81 (2009).
25. C.-H. Lin and Y. Kuo, *MRS Proc.,* 1250-G01-08 (2010).
26. C.-H. Lin and Y. Kuo, *ECS Trans.,* **16**(5) 309 (2008).
27. C.-H. Yang, Y. Kuo, and C.-H. Lin, *Appl. Phys. Lett.* **96**, 192106 (2010).
28. C.-H. Yang, Y. Kuo, C.-H. Lin, and W. Kuo, *MRS Proc.,* 1160-H02-01 (2009).
29. C.-H. Yang, Y. Kuo, C.-H. Lin, and W. Kuo, *Electrochem. Solid-State Lett.,* **14**(1), H50 (2011).
30. C.-H. Yang, Y. Kuo, C.-H. Lin, and W. Kuo, *ECS Trans.,* **33**(3) 307 (2010).
31. C.-H. Yang, Y. Kuo, R. Wan, C. H. Lin and W. Kuo, *MRS Proc.,* **1071**, F02 (2008).
32. H. Nominanda and Y. Kuo, *Electrochem. Solid-State Lett.,* **10**(8) H232 (2007).
33. D. Kahng and S. M. Sze, *Bell Syst. Tech. J.,* **46** 1283 (1967).
34. Y. Kuo and M. Coan, *AM-FPD '09 Proc.,* 259 (2009).
35. M. J. Powell, C. van Berkel, I. D. French, and D.H. Nichols, *Appl. Phys. Lett.* **54**, 1323 (1989).
36. Y. Kuo and M. Coan, *MRS Proc.,* 1245-A12-01 (2010).
37. Y. Kuo, *Electrochem. Solid-State Lett.,* **13**(12) H460 (2010).
38. Y. Kuo, *ECS Trans.,* **33**(5) 357 (2010).
39. W. B. Jackson, C. C. Tsai, and R. Thompson, *Phys. Rev. Lett.,* **64**, 56 (1990).
40. R. E. I. Schropp and J. F. Verwey, *Appl. Phys. Lett.,* **50**, 26(1987).
41. Y. Kuo, *Thin Film Transistors, Materials and Processes, Volume 1: Amorphous Silicon ThinFilm Transistors*, New York, Kluwer, pp. 83, 20, 32(2004).
42. I. Ay and H. Tolunay, Sol. Ener. Mat. Sol. Cells **80**, 209 (2003).
43. B.I. Shklovskii, E.I. Levin, H. Fritzsche, and S.D.Baranovskii, *Transport Correlation and Structural Defects*, ed. H.Fritzsche, World Scientific, pp.161-191 (1990).

ECS Transactions, 37 (1) 167-178 (2011)
10.1149/1.3600737 ©The Electrochemical Society

Understanding the Switching Mechanism in RRAM Devices and the Dielectric Breakdown of Ultrathin High-k Gate Stacks from First Principles Calculations

B. Magyari-Köpe. S.-G. Park, H.D. Lee and Y. Nishi

Department of Electrical Engineering, Stanford University, Stanford, California 94305, USA

RRAM devices received increased interest lately as advanced non-volatile memory technologies in terms of low operating power, high density, better non-volatility, fast switching speed, and compatibility with conventional CMOS process. However, up to date the fundamental physical principles controlling the switching are not well understood. We have employed first-principles simulations based on density functional theory (DFT) to elucidate the effect of oxygen vacancy defects on the electronic structure of rutile TiO_2 and NiO using the local density approximation with correction of on-site Coulomb interactions (LDA+U). The vacancy filament induces several defect states within the band gap, which can lead to the defect-assisted electron transport and account for on-state low resistance conduction in bulk rutile TiO_2 and NiO.
For CMOS devices on the other hand the reliability of the gate stack is becoming a significant challenge with the continuous scaling of transistors, due to the ultrathin oxides and defects in the gate stack. The degradation of the gate oxides has been observed under electrical stress, due to traps generated by defects, e.g. oxygen vacancies present in these materials. First principles methods based on density functional theory are used to determine the location of the defect states in the band gap when these defects are at the various interfaces of the gate stack and how they contribute to the oxide breakdown in ultrathin gate stacks.

Introduction

The technology evolution in the last decade in ubiquitous communications and computing based upon mobile handsets has changed the semiconductor industry with respect to memory-bit consumption and memory usage. Especially looking into 3G and beyond which supports multimedia functions such as digital photography, MP3, video on demand, it is apparent that a significantly larger number of bits of memory are required on a single chip. Embedded memory may consume as much as 90% or more of a typical chip of this nature. Power consumed by such memory areas of the chip including in the active and stand-by states will quickly become a substantial portion of the entire power budget.
The nanoparticle-based memory is to replace floating gate, resistance change memory, phase change memory, ferroelectric memory, magnetic spin based memory and even organic/molecular memory are being pursued. The development of these new non-volatile memories, needs new materials that go far beyond the common practice of the traditional semiconductor industry and technology.
Resistive switching devices based on either reversible filamentary percolation path, bulk defect/trap conduction or interface modifications represent a high-risk, high-payoff approach to embedded non-volatile memory (NVM). Switching devices that use a resistive change in a metal-insulator-metal (MIM) stack for non-volatile storage starts to show up on the ITRS emerging technologies roadmap. There appears to be two broad

classes of devices, one based on solid ionic conductors and the other based on the transition metal oxides. Resistive switching materials offer a low cost, low temperature, non-volatile memory module that may be compatible with back-end processing. They offer the possibility for highly scalable, low power operation with non-destructive readout and even multiple bits per cell. But little is known about the fundamental switching mechanisms, so it is difficult to gauge reliability, endurance, retention or even the true scaling limits of the devices.

The interest in the complex transition metal doped oxides emerged from research on high temperature superconductivity [1-4] and DRAM capacitor dielectrics [5,6]. Simple binary metallic oxides (NiO, Cu_2O) also exhibit switching, first observed in the early 1960's [7,8], and rediscovered along with the metal sulfides as a potentially scalable memory solution [9-11]. Filamentary mechanisms are thought to be at work, though the influence of simultaneous electronic and ionic conduction through lattice vacancies may also contribute to the fundamental mechanism of switching [12]. In recent years, NiO, HfO_x, AlO_x, TiO_x has been intensively studied. Continuous exploration for the use of these RRAM materials and study for the interaction between the RRAM materials with the electrode materials (e.g. Pt, Ti, TiN, W, Al, Cu, etc.) are in progress.

In this paper we discuss the fundamental aspects of the switching mechanism based on both theoretical and experimental characterization of several RRAM materials. The filament formation model and its implications for "ON" state conduction mechanism in TiO_2 and NiO is established based on first principles calculations.

For the MOSFETs, during the past years considerable progress has been achieved in the development of high-k based MOSFET technology. In the recent years, experimental and theoretical efforts have been directed towards characterizing a large number of metals on high-k dielectric materials, and towards understanding the fundamental processes undergoing at the various types of interfaces in the gate stack. The interfacial properties generally include the physical alignment/misalignment at the atomic level induced during deposition, e.g. dangling bond generation, coordination number change, metal surface orientation, all these leading to detectable charge transfer in the metal-oxide interface layer and ultimately to the formation of an electric dipole that causes the metal work function variation. Another contribution to the interfacial chemistry is from the presence of defects and impurities segregated at the interface. These point defects act as traps and can lead to flat band voltage instabilities. Currently, experiments point to the fact that most probably the oxygen vacancies are the main source of device degradation. As the thickness of the gate dielectric is further reduced to a few nanometers, and the gate dielectric is under constant voltage or current stress, further traps due to defects are generated and can form a conductive path. Up to date, however, we are lacking a detailed understanding and characterization of the type of oxygen vacancies that are most predominantly present in the gate stack and how they affect the device performances. Here we apply *ab initio* techniques to address these critical issues in the design and fabrication of new transistor gate stack structures. Atomistic calculations have been carried out for a model system configuration of the entire gate stack, metal-HfO_2-SiO_2-Si. The effect of oxygen vacancies positioned in the vicinity of the metal-HfO_2, HfO_2-SiO_2 and SiO_2-Si interfaces and in the bulk region of the HfO_2 layer on the structure, electronic structure, and band offsets were determined.

1. Computational Details of the First Principles Modeling

The structures and energies of oxygen deficient rutile TiO_2 and cubic NaCl-type NiO were calculated using density functional theory. In the past decade several theoretical investigations employed the local density (LDA) and the gradient corrected approximations (GGA), to calculate the electronic structure of transition metal oxides [13-15]. It was found necessary to go beyond LDA and GGA however, to correct for some of the most severe shortcomings of the conventional LDA and GGA, i.e. the accurate prediction of the energy band gap and the position of the electronic defect states in the band gap. Recent theoretical developments, as the addition of the on site Coulomb correction within LDA/GGA+U [16], dynamical mean field theory approaches [17], the progress in hybrid functional [18] and GW implementations [19], have been very successful in predicting realistic properties for these materials.

In this study the electronic interactions are described within the LDA/GGA+U formalism for TiO_2 and NiO, where on-site Coulomb corrections are applied on the $3d$ orbital electrons of Ti or Ni atoms (U^d) for TiO_2 and NiO respectively, and also on the $2p$ orbital electrons of the O atoms (U^p), in the case of TiO_2. While for NiO the GGA+U^d approach yields an acceptable band structure, for TiO_2 this method is not adequate. The U^d parameter is shown to affect only the character of the conduction band and values higher than 8 eV produce an unphysical description of the electronic interactions in TiO_2. The results are dramatically improved, when correlation corrections are introduced additionally on the O $2p$ orbitals by employing the LDA+U^d+U^p approach, and we observe systematic shifts for both the valence and conduction bands. This combined approach produces a corrected energy band structure and the band gap energy is in very good agreement with experimental data for TiO_2. For the oxygen deficient structure, in the case of a neutral oxygen vacancy we have shown that the electrons from the Ti nearest to the vacancy become localized and induce defect states within the band gap, in very good agreement with experiment and other theoretical calculations. Additionally, the electronic states are strongly localized on the three Ti atoms surrounding oxygen vacancy. The doubly positively charged oxygen vacancy was found to be stable over the neutral oxygen vacancy in a large range of Fermi energy values and the vacancy formation energies show significant dependence on the deposition conditions, i.e. Ti or O rich environments. For the CMOS structure, we have considered TiN-HfO2-SiO2-Si-Pt structure, with 2nm of HfO2, 1nm of SiO2 and 1.2 nm of Si. The metal thicknesses on both sides were 1.5 nm. Atomically sharp interfaces had been created between each material, and all atoms were subsequently relaxed to their equilibrium position. We have constructed a database of structures, which includes oxygen deficient and excess configurations to search for the most stable configuration. The effect of oxygen vacancies positioned in the vicinity of the metal-HfO2, HfO2-SiO2 and SiO2-Si interfaces and in the bulk region of the HfO2 layer on the structure, electronic structure, and band offsets were determined.

The density functional calculations were done using the Vienna *ab initio* program package, (VASP) [20-22], and the projector augmented-wave pseudopotentials (PAW). An energy cutoff of 353 eV for TiO_2, and 500 eV for NiO was employed for the plane wave expansion. For k-point integration, a Monkhorst-Pack grid of 8x8x12 grid for the TiO_2 primitive cell, 4x4x4 for the $Ti_{72}O_{144}$ supercell, and a 2x2x2 grid for the supercell of $Ni_{64}O_{64}$ was used. For the gate-stack structure a 4x4x1 grid had been used for the supercell of $Ti_{16}N_{16}Hf_{18}Si_{25}O_{50}Pt_{28}$. The O $2s^2 2p^4$, Ti $3s^2 3p^6 3d^2 4s^2$, Ni $3d^8 4s^2$, Hf $5p^5 d^2 6s^2$, Si $2s^2 2p^2$, N $2s^2 2p^3$ and the Pt $5d^9 6s^1$ states were considered valence electrons.

All atoms were allowed to relax with energy convergence tolerance of 10^{-6} eV/atom and ground state was obtained by minimizing the force on each atom to be less than 0.01 eV/Å.

2. Modeling of the Filamentary Structures in RRAM Devices

2.1. Overview of the Forming and Switching Mechanisms

Earlier models discuss the importance of oxygen vacancies to achieve resistive switching, e.g. oxygen vacancies have been shown to induce a metal-insulator transition in $SrTiO_3$ (STO) single crystals [23]. The aggregation of these dopants into an extended defect network is postulated to drive the generation of conductive filaments during resistive switching [24]. The model for electroforming suggests that molecular oxygen will evolve according to the reaction

$$O_O \longleftrightarrow \tfrac{1}{2} O_2 (g) + V_O^{**} + 2e^-$$

where, in Kröger-Vink notation, O_O indicates an oxygen ion on a lattice site and V_O^{**} represents a doubly charged oxygen vacancy.

The electroforming step is often required before bistable switching can be observed in RRAM devices. Janousch, et al. experimentally observed in lateral devices on the surface of SrTiO3 single crystals that a channel of oxygen vacancies (V_O) existed after electroforming [25].

In stoichiometric TiO_2, after the removal of an oxygen ion, the oxygen vacancy is found to behave as a positively charged double donor [26, 27]. Both the crystalline and thin film reduced TiO_{2-x}, displays substantially higher n-type conductivity than its stoichiometric counterpart. Reduced TiO_2, corresponds to several stable titanium-oxygen phases that exist between Ti_2O_3 and TiO_2, in the phase diagram of the Ti-O system. The vacancy formation mechanisms had been studied by Yang et. al. [28] in Pt/TiO2/Pt devices and found that the positive electroforming voltage to the top electrode causes bubble formation. The observed deformation is attributed to the evolution of oxygen gas formed from the discharge of O^{2-} ions that drifts toward the anode, while the V_O are drawn to the cathode and decrease the field in this region, thus 'freezing' the vacancies in place [24].

As for switching models, Jameson et al. [29], and Dong et al. [30], identified that oxygen vacancies were responsible for TiO_2 bipolar switching through field-driven drift, which would alter the Schottky barriers at the electrode interfaces. Further studies are necessary to understand the details of the switching behavior: the role of both interfaces and the interface interactions with reactive electrodes [31].

For nanoscale filaments inside vertical Pt/TiO2/Pt devices, in a recent work, Kwon et al. [32] identified the filaments to be of Ti_nO_{2n-1} composition, known as room temperature conducting Magnéli phases. Using high-resolution transmission electron microscopy (HRTEM) and electron diffraction they observed that the filaments form a bridge between the two electrodes in the SET state. Direct probing of the filaments using conductive atomic force microscopy revealed that the filaments were both localized and conducting and the character changed abruptly from metallic to semiconducting near 130 K. Conversely, after RESET, no Magnéli phases were observed in the regions previously occupied by conducting filaments. Thus, TiO_2 is believed to spontaneously order into Ti_4O_7 when the concentration of oxygen vacancies reaches a critical density.

The NiO-based RRAM has been extensively investigated due to its unipolar switching characteristics. Out of the proposed models to explain the unipolar switching phenomena, the "filament model" has been found to give qualitative explanation for unipolar switching [33]. To understand in detail the mechanism, several considerations associated with switching phenomena had to be taken into account, i.e. migration of oxygen into Pt anodic electrode after the forming process [34], metallic nickel defects in NiO [35], oxygen migration [36], thermal energy considerations [7,37], crystal disorder and electrode interface effects [38-40]. An atomistic description of the filament has been recently proposed [41], and the role of the oxygen vacancies for switching was investigated, then a theoretical model was established to explain the formation/rupture of a metallic filament in NiO.

2.2. Modeling the Filament Properties in TiO_2

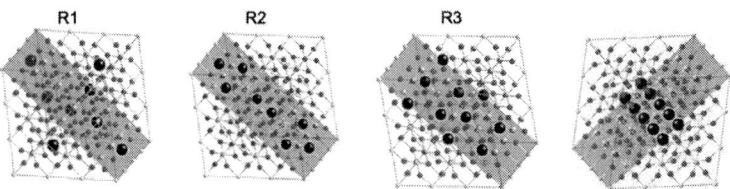

Figure 1. The schematic picture of the R1, R2 and R3 random configurations and the [001] oxygen vacancy ordering system in 3x3x4 supercell of rutile TiO_2. Red spheres represent oxygen, blue spheres represent Ti, and the black spheres represent oxygen vacancies. All oxygen vacancies are on the same (110) plane.

TiO_2 has been found to exhibit n-type semiconducting behavior due to extra electrons generated by intrinsic defects such as oxygen vacancies [50,51]. The oxygen vacancy induces a defect state in the band gap, and the semiconducting behavior depends on the position of the defect state relative to the conduction band minimum (CBM). Several recent theoretical calculations were reported that try to elucidate the controversy around the position of the oxygen vacancy defect state in reduced TiO_2 [52-54]. The differences in the description of the position of defect state by different methods arouse because there is a strong correlation effect observed between the Ti and O atoms and the level of accuracy describing that is crucial to get reasonable results. Based on density functional calculations employing the LDA+U method we have previously shown that an isolated oxygen vacancy induces a defect state at 0.4 eV below the CBM [27]. Thus, the oxygen vacancy defect states are too deep to elevate electrons to the conduction band at room temperature. In addition, those electrons were found to be strongly localized on both the oxygen vacancy site and surrounding Ti. Therefore it is widely accepted now that the isolated mono oxygen vacancy is not likely to produce the n-type semiconductivity observed in this material. Here, we explore different configurations of oxygen vacancies shown in Figure 1, three random arrangements of oxygen vacancies, and an ordered vacancy cluster, and investigate what is the effect on the conduction properties of reduced TiO_2. Previously, Szot et al., [55] have examined the effect of dislocation with oxygen vacancies on the conduction properties in $SrTiO_3$, and more recently it was experimentally observed that the switching between an insulating and conductive TiO_{2-x} is associated with a phase transition from rutile to Magnéli phase in a filament with metallic conduction behavior [32]. Therefore, our investigation is directed towards

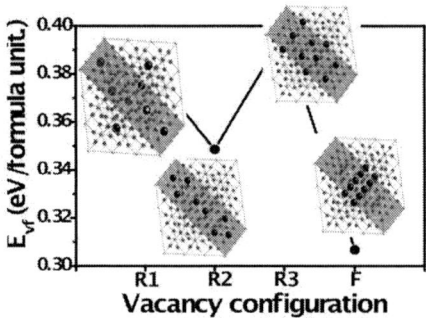

Vacancy configuration

Figure 2. Total energy of the [001] ordered vacancy structure compared to the total energy of the random configurations from Figure 1.

characterizing vacancy-ordering mechanisms, in which certain configurations are found to exhibit higher metallic conducting properties. The thermodynamic stability of this ordered vacancy structure has been calculated and compared with that of randomly distributed vacancy configurations. The vacancy formation energy was determined using the following formula:

$$E_{vf} = E(TiO_{2-x}) - E(TiO_2) + n\mu_O$$

where $E(TiO_{2-x})$ is the total energy of a supercell containing the oxygen vacancy, $E(TiO_2)$ is the total energy of a perfect TiO_2 in the same size of supercell, μ_O is the oxygen chemical potential, and n is the number of oxygen vacancies.

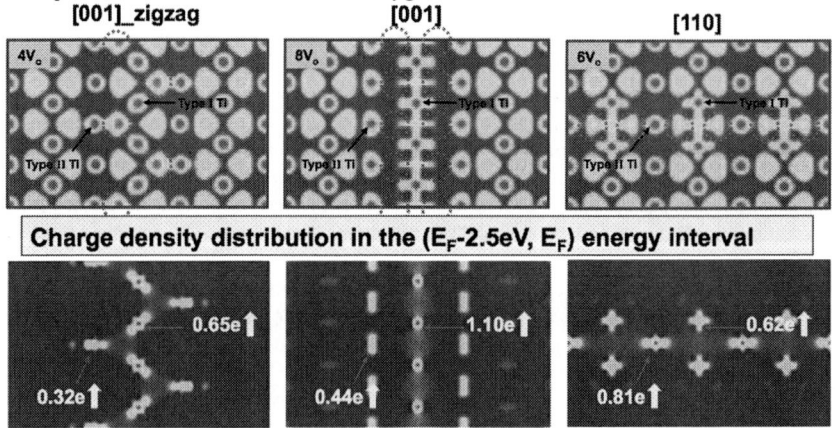

Figure 3. Electron localization function and charge density distribution in the window of (E_F-2.5eV, E_F) of (a) zig-zag type of ordering of $4V_O$/simulation cell along the [001] direction; (b) full chain with $8V_O$/simulation cell along the [001] direction; (c) chain of $6V_O$/simulation cell along the [110] direction of the (001) plane.

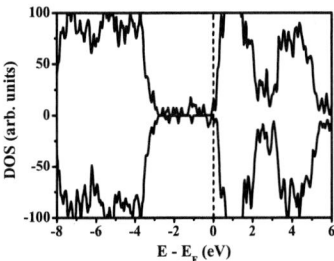

Figure 4. Total density of states (TDOS) of the [001] ordered vacancy structure;

From Fig. 2, the ordered vacancy structure has the lowest vacancy formation energy relative to a number of randomly configured vacancy structures of percolative nature. The lowest energy structure corresponds to the ordered vacancy configuration composed of two parallel vacancy chains along the [001] direction, shown in Fig. 3. The character of the atomic bonding around the vacancies can be discussed by inspecting the electron localization function and charge density distribution of partially and fully ordered vacancy configuration structure along the [001] direction, as shown in Fig. 3. While in the partial ordering regime in a zig-zag arrangement of oxygen vacancies (Fig. 3 (a)) and the oxygen vacancy chain structure along the [110] direction (Fig. 3 (c)) some degree of delocalization of electrons are observed around the vacancy sites, however in the case of a full chain of ordered vacancies the electron localization on the vacancy sites becomes negligible, thus most electrons localize on the Ti ions in between the two vacancy chains, forming a metallic Ti chain (Fig. 3 (b)). This results in Ti-Ti bonds involving a strong overlap of t_{2g}-t_{2g} type orbitals that are responsible for the metallic conductive channel through the Ti ions along the [001] direction. The total density of states (TDOS), shown in Fig. 4 for the vacancy ordered structure along [001] direction provides evidence that several discrete defect states are created and distributed throughout the band gap with a partial overlap between them. The iso-surface of the band decomposed charge density of all defect states within the band gap for the random and ordered oxygen vacancy configurations presented in Fig 1, are shown in Fig. 5. We conclude that the oxygen vacancies act as mediators of the electron conduction, while the actual electron conduction is through the successive Ti ions in the channel. Therefore, we propose that the conduction mechanism in RRAM devices based on TiO_2 is likely to be a defect assisted tunneling through metallic Ti ions, not of pure electron doping by shallow defect levels. Most of the defect states originate from the centered Ti ions in between the two vacancy chains, and the atomic charge density around those Ti ions are considerably increased compared to the stoichiometric TiO_2, supported by the electron charge density distribution plots (Fig. 3 and Fig. 5).

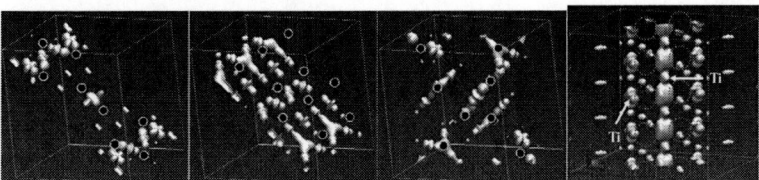

Figure 5. Iso-surface (0.1 e/Å^3) of band decomposed charge density of all defect states for the a) R2 ; b) R3; c) R1 and d) vacancy chain structure.

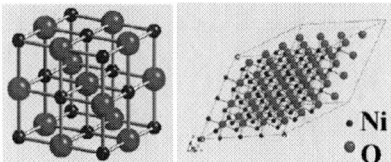

Figure 6. (a) Unit cell of NiO in simple-cubic NaCl structure; (b) Supercell of $Ni_{64}O_{64}$ used in the calculation for oxygen vacancy migration.

2. 3. Modeling the Filament Properties in NiO

NiO had been proposed to have p-type conductivity. In order to understand its nature, in this study, isolated oxygen and nickel vacancies had been introduced in the supercell $Ni_{64}O_{64}$, shown in Fig. 6, and the formation energy was calculated. We have found that the oxygen vacancy formation is favorable due to its low formation energy and the stable charge state of oxygen vacancy is +2. The as-deposited NiO has been found nickel deficient. We show that the vacancy states of charged nickel vacancies are stable for E_F near the valence band maximum (VBM) as shown in Fig. 7. The transition states of nickel vacancies can be determined in Figure 7(a) at which the charged Ni vacancies become stable, -0.04 eV for q=-1 and 0.16 eV for q=-2. These low energies for acceptor-like states support the possibility of p-type conductivity in Ni-deficient NiO films observed in experiments [56]. However, after forming process, oxygen vacancies are generated [34] and clustered due to the interaction between oxygen vacancies [57] as shown in Fig. 8. With more oxygen vacancies, E_F moves from near VBM to the level above midgap, which implies that the material property at local conductive region changes from p-type to n-type.

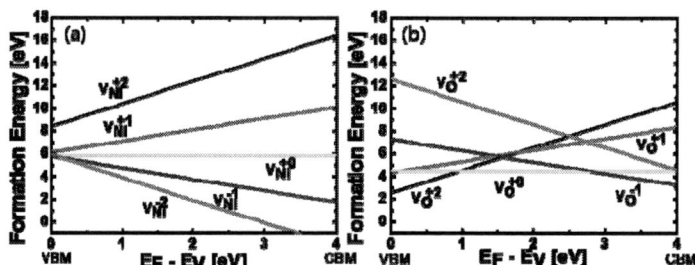

Figure 7. – Formation energy of charged vacancies (a) for V_{Ni}, (b) for V_O.

Figure 8. shows the theoretical analysis of the atomic structure for the "ON" state conduction, i.e. the interaction between 6 aligned oxygen vacancies in the <110> direction and for the "OFF" state we assume that some of these oxygen vacancy sites become occupied by oxygen atoms [41]. The calculated total density of states (DOS) (Fig. 8(a)) shows that the "ON" state has metallic property. The density of states near E_F for the proposed "ON" state configuration is dominated by the 4 metallic nickel atoms surrounded by oxygen vacancies having 9.8 electrons localized near their sites, resulting in metallic conduction through the metallic nickel atoms. All the states near or below

Fermi level come from vacancy neighboring nickel atoms, which are forming a filament. The transport mechanism is proposed to be of polaron hopping type, and this effect is in accord with experimental results of Jung et al [58]; where the high resistance

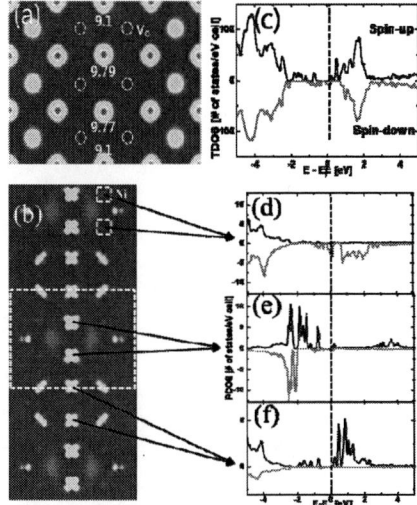

Figure 8. (a) Charge density of supercell with six oxygen vacancies and (b) partial charge density within (E_F, E_F + 0.3eV) in (100) plane including oxygen vacancies and Ni metal chain. (c) Total density of states of the supercell and (d)-(f) partial density of states of d orbitals at each Ni atom. Fermi level is set to 0 eV (c)-(f).

state was supposed to have both properties of weak metallic conduction and polaron hopping. The states just above Fermi level and below CBM correspond tonickel atoms in the filament (Figures 8(b) and 8(d)–8(f)).. Figure 8(b) shows the band decomposed (partial) charge density within (E_F, E_F + 0.3 eV) suggesting that the transport path is in a direction of filament. The ordered oxygen vacancy configurations induce the redistribution of electrons around metallic type Ni chains, which can be regarded as evidence of filamentary conduction in NiO-based resistive switching devices.

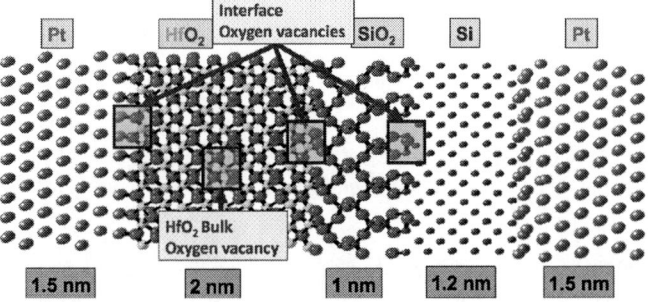

Figure 9. – Schematic illustration of the position of defect states along a gate stack composed of 1.5 nm TiN – 2nm HfO_2 -1 nm SiO_2 – 1.2 nm Si -1.5 nm Pt.

2. Modeling of the Interface and Bulk Defects in CMOS Devices

The oxygen defects are placed in the bulk region of HfO_2 and at the metal-oxide, oxide-oxide and oxide-semiconductor interfaces as shown schematically in Figure 9 and in more detail in Figure 10. The structures containing the defects had been optimized and energy gap values of 5.2 for HfO_2, 8 eV for SiO_2 and 1.1 eV for Si had been obtained using the GGA+U method. The partial density of states for these structures are shown in Figure 11, corresponding to a structure with no defects, one with bulk HfO_2 vacancy and o ne vacancy at the HfO_2-SiO_2 interface.

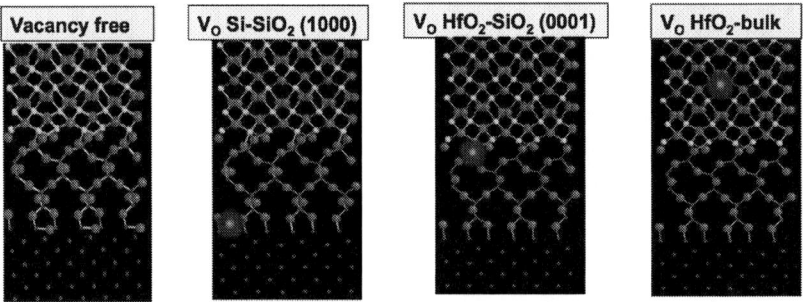

Figure 10. –The atomic structure of the simulated gate stack showing the position of defect states at the HfO_2 - SiO_2, Si-SiO_2 interfaces and in the HfO_2 bulk region.

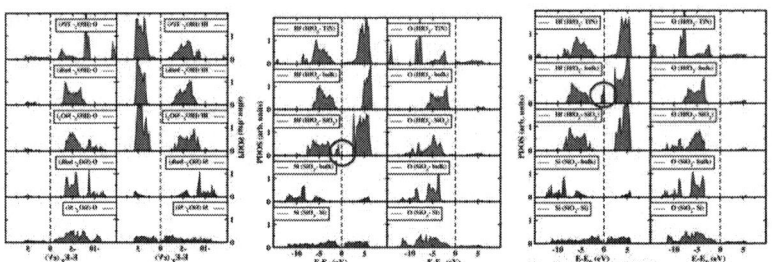

Figure 11. – Partial density of states for Hf, Si and O in different regions of the gate stack, in the neighborhood of a vacancy and further away from a vacancy site.

From the partial density of states shown in Figure 11, lower lying defects states (close to the VBM) are observed for vacancies at the HfO_2-SiO_2 interface, while the bulk defect states are mid gap positioned. Therefore, for implications on the breakdown of the oxide, the oxide-oxide interfaces will have effects on the I-V characteristics at lower voltages than the bulk vacancies in the high-k oxide.

Conclusions

The "ON"-state implications of vacancy ordering on the conduction mechanism were discussed based on first-principle calculations for rutile TiO2 and NiO with oxygen vacancies. The conductive channel is formed between nearest Ti and Ni ions, respectively and the oxygen vacancy chains mediate the conduction. When the oxygen vacancy concentration is increased and the vacancies become nearest neighbors significant charge redistribution is observed, resulting in the enhanced metallic nature of the interaction between nearest Ti and Ni ions, respectively. Therefore, the electrons transport can be described as defect assisted tunneling through the channel of Ti and Ni ions in these binary metal oxides.

For the CMOS structures, oxygen vacancy defects in different regions of the gate stacks have different effects on the density of states, thus on the I-V characteristics, depending on the relative position of the defect state they create in the band gap. The strongest contribution is coming from the vacancies at the HfO_2-SiO_2 interfaces, because their effect will be observed at lower voltages, thus the dielectric breakdown may originate from this interface, and since the Si-SiO_2 interface defects do not show strong electronic signature.

Acknowledgement

This work was sponsored by the Stanford NMTRI, and INMP center and the MSD program of the Marco Focus Center. M. T. was supported by an SRC/Intel fellowship. The computational work was carried out using the National Nanotechnology Infrastructure Network's Computational Cluster at Stanford.

References

[1] S.Q. Liu at al., Appl. Phys. Lett., **76**, 2749, (2000).
[2] A. Baikalov et al., Appl. Phys. Lett., **83**, 957, (2003).
[3] S. Tsui et al., Appl. Phys. Lett., **85**, 317, (2004).
[4] S. Duhalde et al., Physica B, **354**, 11, (2004).
[5] A. Beck et al., Appl. Phys. Lett., **77**, 139, (2000).
[6] Y. Watanabe et al., Appl. Phys. Lett., **78**, 3738, (2001).
[7] J.F. Gibbons and W.E. Beadle, Solid State Electr., **7**, 785, (1964).
[8] D.C. Bullock and D.J. Epstein, Appl. Phys. Lett., **17**, 199, (1970).
[9] S. Seo et al., Appl. Phys. Lett., **85**, 5655, (2004).
[10] S. Seo et al., Appl. Phys. Lett., **86**, 093509, (2005).
[11] T. Sakamoto et al., Appl. Phys. Lett., **82**, 3032, (2003).
[12] R. Waser and M. Aono, Nat. Mater. **6**, 833, (2007).
[13] S. B. Sinnott et al., Phys. Rev. B **61**, 15645, (2000).
[14] K. M. Glassford and J. R. Chelikowsky, Phys. Rev. B **46**, 1284, (1992).
[15] C. Lee et al., Phys. Rev. B **50**, 13379, (1994).
[16] S. L. Dudarev et al., Phys. Rev. B **57**, 1505, (1998).
[17] A. I. Liechtenstein et al. , Phys. Rev. B **52**, R5467, (1995).
[18] J. Heyd et al., J. Chem. Phys. **118**, 8207, (2003).
[19] S. V. Faleev et al., Phys. Rev. Lett. **93**, 126406, (2004).
[20] G. Kresse and J. Furthmüller, Phys. Rev. B **54**, 11169 (1996); Comput. Mater. Sci. **6**,

15 (1996).

[21] P. E. Blöchl, Phys. Rev. B **50**, 17953 (1994).

[22] G. Kresse and D. Joubert, Phys. Rev. B **59**, 1758 (1999).

[23] K. Szot, et al., Phys. Rev. Lett., **88**, 075508, (2002).

[24] R. Waser, et al., Adv. Mater., **21**, 2632 (2009).

[25] M. Janousch, et al., Adv. Mater. **19**, 2232 (2007).

[26] B. Poumellec, et al., J. Phys. Condens. Matter, **3**, 8195, (1991).

[27] S.-G. Park, et al., Phys. Rev. B, **82**, 115109, (2010).

[28] J.J. Yang, et al., Nanotech. **20**, 215201, (2009).

[29] J.R. Jameson, et al., Appl. Phys. Lett. **91**, 112101, (2007).

[30] R. Dong, et al., Appl. Phys. A, **93**, 409, (2008).

[31] H.Y. Jeong, et al., Adv. Funct. Mater. (2010).

[32] D.-H. Kwon, et al., Nature Nanotech., **5**, 148 (2010).

[33] B. J. Choi, et al., J. Appl. Phys. **98**, 033715, (2005)

[34] M.-J. Lee, et al., Nano Lett. **9**, 1476, (2009)

[35] J. G. Simmons and R. R. Verderber, Proc. R. Soc. London, Ser. A **301**, 77, (1967)

[36] H. Tang, et al., Appl. Phys. Lett. **71**, 2560, (1997)

[37] Y. Sato, et al., Appl. Phys. Lett. **90**, 033503, (2007)

[38] C. Park, et al., Appl. Phys. Lett. **93**, 042102, (2008)

[39] C. B. Lee, et al., Appl. Phys. Lett. **93**, 042115, (2008)

[40] S. R. Lee, et al., Jpn. J. Appl. Phys. **49**, 031102, (2010)

[41] H.D. Lee, et al., Phys. Rev. B, **81**, 193202, (2010)

[42] J.P. Joshi, et al., J. Phys.: Cond. Matt. **16**, 2869, (2004).

[43] A.-M. Haghiri Gosnet, et al., J. Phys. D: Appl. Phys. **36**, R127, (2003).

[44] C. Jooss, et al., Proc. Nat. Acad. of Sci. **104**, 13597, (2007).

[45] Y.B. Nian, et al., Phys. Rev. Lett. **98**, 146403, (2007).

[46] A. Sawa, et al., Appl. Phys. Lett. **85**, 4073, (2004).

[47] R. Fors, et al., Phys. Rev. B **71**, 045305, (2005).

[48] Z.L. Liao, et al., Appl. Phys. Lett. **94**, 253503, (2009).

[49] D.-J. Seong, et al., IEDM Tech. Dig. (2009).

[50] M. D. Earle, Phys. Rev. **61**, 56, (1942).

[51] D. C. Cronemeyer, Phys. Rev. **113**, 1222, (1959).

[52] J. Chen, et al., Phys. Chem. Solids **62**, 1257, (2001).

[53] E. Cho, et al., Phys. Rev. B **73**, 193202, (2006).

[54] M. Ramamorthy, et al., Phys. Rev. B **49**, 7709, (1994).

[55] K. Szot, et al., Nat. Mater. **5**, 312, (2006).

[56] A. J. Bosman, et al., Phys. Lett. **19**, 372, (1965).

[57] S. Park, et al., Phys. Rev. B 77, 134103, (2008).

[58] K. Jung, et al., Appl. Phys. Lett. **90**, 052104, (2007).

CHAPTER 5

CHALLENGES IN PROCESS ENGINEERING

ECS Transactions, 37 (1) 181-188 (2011)
10.1149/1.3600738 ©The Electrochemical Society

Atomically Controlled CVD Processing for Doping
in Future Si-Based Devices

Junichi Murota[1,*], Masao Sakuraba[1] and Bernd Tillack[2,3]

[1] Laboratory for Nanoelectronics and Spintronics, Research Institute of Electrical Communication, Tohoku University, 2-1-1 Katahira, Aoba-ku, Sendai 980-8577, Japan
[2] IHP, Im Technologiepark 25, 15236, Frankfurt (Oder), Germany
[3] Technische Universität Berlin, HFT4, Einsteinufer 25, 10587, Berlin, Germany
* Tel:+81-22-217-5548, Fax:+81-22-217-5551 E-mail: murota@riec.tohoku.ac.jp

By atomic layer formation of B or P on $Si_{1-x}Ge_x$ (100) surface and subsequent Si capping layer deposition, heavy atomic-layer doping is achieved at temperatures below 500 °C. B doping dose of about 7×10^{14} cm^{-2} is confined within an about 1 nm thick region, but the sheet carrier concentration is as low as 1.7×10^{13} cm^{-2}. The in-situ B doping in tensile-strained Si epitaxial growth suggests that the low electrical activity is caused by B clustering as well as the increase of interstitial B atoms. For unstrained Si cap layer grown on top of the P atomic layer formed on $Si_{1-x}Ge_x$(100) with P atom amount below about 4×10^{14} cm^{-2} using Si_2H_6 instead of SiH_4, it is found that tensile-strain in the Si cap layer growth enhances P surface segregation and reduces the incorporated P amount around the heterointerface. The electrical inactive P atoms are generated by tensile-strain in heavy P doped region. These results demonstrate that atomically controlled processing for doping is influenced by strain.

Introduction

Atomically controlled processing has become indispensable for the fabrication of Si-based ultrasmall devices and Si-based heterodevices for ultra-large-scale integration, because high performance devices require atomic order abrupt heterointerfaces and doping profiles as well as strain engineering due to introduction of Ge into Si. Our concept of atomically controlled processing is based on atomic-order surface reaction control (1-3). The final goal is the generalization of atomic-order surface reaction processes and the creation of new properties in Si-based ultimate small structures which will lead to nanometer scale Si devices as well as Si-based quantum devices (**Fig. 1**).

In our previous work, high-quality low-temperature epitaxial growth of Si, Ge and $Si_{1-x}Ge_x$ with atomically flat surface and interfaces on Si(100) was achieved by ultraclean low-pressure CVD, and in-situ doped $Si_{1-x}Ge_x$ epitaxial growth on the (100) surface in a SiH_4-GeH_4-dopant (PH_3 or B_2H_6 or SiH_3CH_3)-H_2 gas mixture. The deposition rate, the Ge fraction and the dopant concentration have been expressed quantitatively by modified Langmuir-type rate equation (2). Self-limiting formation of 1-3 atomic layers of group IV or related atoms in the thermal adsorption and reaction of hydride gases (SiH_4, GeH_4, NH_3, PH_3, B_2H_6 CH_4 and SiH_3CH_3) on Si(100) and Ge(100) were generalized based on the Langmuir-type model (3). In many case, hydride molecules are adsorbed and react

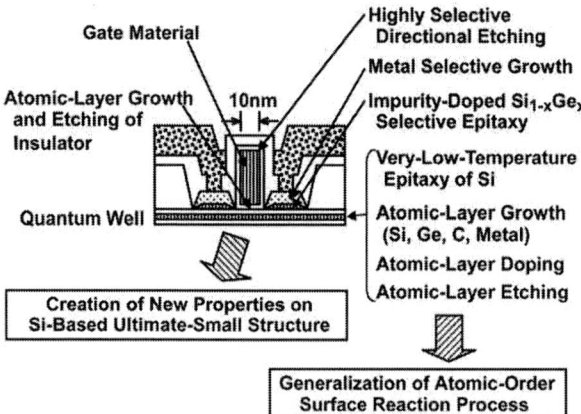

Figure 1 Atomically controlled processing for group IV semiconductors for ultrasmall and nanodevices.

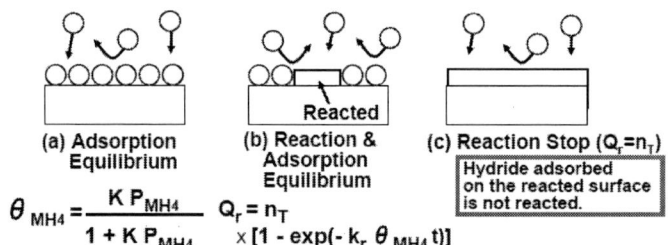

Figure 2 Schematic image of self-limited reaction of hydride for atomic-order growth based on Langmuir-type model.

simultaneously on the surface as shown in **Fig. 2** (3). Moreover, atomic-layer doping was performed by Si or $Si_{1-x}Ge_x$ epitaxial growth on N, P, B or C atomic layer already-formed on Si(100) or $Si_{1-x}Ge_x$ (100) surface (2-5).

In this paper, we review heavy atomic-layer doping of B and P by atomic layer formation of B or P on Si or $Si_{1-x}Ge_x$ (100) surface and subsequent Si capping layer deposition. Moreover, the influences of strain in Si epitaxial growth on the doping characteristics are described.

Boron Atomic-Layer Doping in Si Epitaxial Growth

Surface Reaction of B_2H_6 on Si(100)

As shown in **Fig. 3** (6), by B_2H_6 exposure on the Si(100) at 180 °C, the B atomic amount tends to saturate self-limitedly at around 1.4×10^{15} cm^{-2} (2 AL). In the case at 500

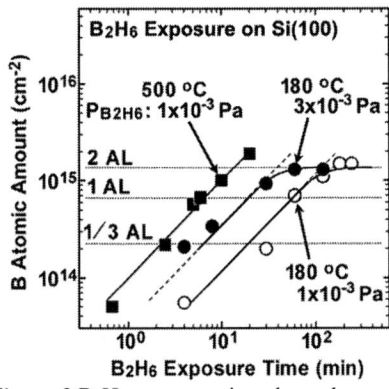

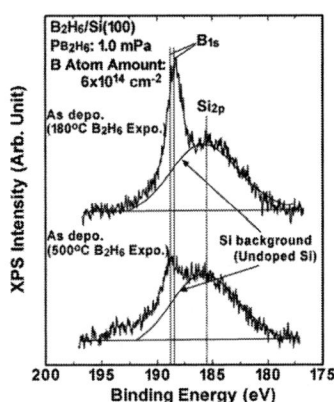

Figure 3 B_2H_6 exposure time dependence of B atomic amount on Si(100) at 180 and 500 °C.

Figure 4 B 1s XPS after air exposure from B atomic-layer on Si(100) formed at 180 °C and 500 °C with B atomic amount of about 6×10^{14} cm^{-2}.

°C, the B atom amount increases with B_2H_6 exposure time and exceeds 2 AL. It is clear that continuous B_2H_6 reaction at 500 °C proceeds with H desorption on B atoms. Typical B 1s XPS spectra obtained from the B AL with B atomic amount with 6×10^{14} cm^{-2} on Si(100) after air exposure (**Fig. 4**) shows that an elemental B peak appears at around 187-190 eV (7) for both the temperatures of 180 °C and 500 °C, while the peak for 500 °C is smaller than that for 180 °C. A B oxide peak at 190-195 eV (7) is observed hardly for 180 °C but for 500 °C. This supposes that H adsorption on B atoms protects B oxidation. Therefore, it is suggested that the H adsorption on the B AL maintains at 180 °C whereas H desorption from the B AL proceeds at 500 °C.

Si Epitaxial Growth over B Atomic Layer already-formed on Si(100) Surface

By Si epitaxial growth with SiH_4 reaction at 500 °C on top of B atomic layer of about 7×10^{14} cm^{-2} formed with B_2H_6 reaction at 180 °C on Si(100) surface, most of the incorporated B atoms are confined within an about 1 nm-thick region as shown in **Fig. 5**. The total B atom amount is in good agreement with the initial B atom amount before the subsequent Si epitaxial growth (5), but the sheet carrier concentration is as low as 1.7×10^{13} cm^{-2}. The electrical inactive B could be a result of clustering of surface B atoms just before Si deposition at 500 °C. It is well-known that B clustering reduces the electrical activity of B in Si. Additionally, by heavy B doping, strain could be induced. For in-situ B-doped Si epitaxial growth on 0.8%-tensile-strained Si(100)-on-insulator (SOI) at 550 °C using SiH_4 and B_2H_6, the B concentration region in which carrier concentration is almost equal to the B concentration is lower in the strained SOI, compared with that on the unstrained SOI shown in **Fig. 6** (8). Additionally it was found that, for in-situ B-doped Si epitaxial growth, the substitutional B and the interstitial B concentrations in the unstrained Si is higher and lower than those on tensile-strained Si, respectively, in other words, substitutional B concentration is higher in the compressive-strained Si, compared with that in the unstrained or tensile-strained Si. These results indicate suppression of B

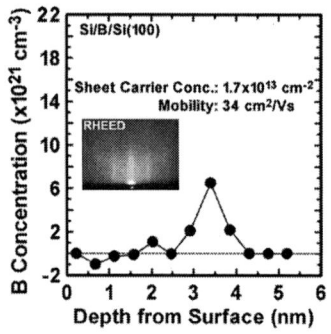

Figure 5 Depth profile of B concentration, carrier concentration and mobility in the B atomic-layer doped Si films on Si(100) without low-temperature SiH$_4$ exposure at 180-300 °C before capping Si deposition at 500 °C. Initial B atomic amount is 7.0×10^{14} cm^{-2}.

Figure 6 Relationship between carrier concentration measured by van der Pauw method and B concentration measured by SIMS in in-situ doped Si formed on unstrained and on 0.8% tensile-strained SOI at 550 °C.

clustering due to low temperature processing as well as enhanced generation of substitutional B due to compressive strain.

Phosphorus Atomic-Layer Doping in Si Based Epitaxial Growth

Surface Reaction of PH$_3$ on Si(100)

In the case of PH$_3$ reaction on Si(100) and Ge(100) (9), it is found that the atomic amount of P layer formed on the surface depends on the PH$_3$ exposure temperature as shown in **Fig. 7**. The PH$_3$ reaction is suppressed on the hydrogen-terminated Si and Ge surfaces, but PH$_3$ is adsorbed dissociatively on the hydrogen-free Si and Ge surfaces at 300 °C and 200 °C, respectively. As a result, the P atomic amount on the surface tends to

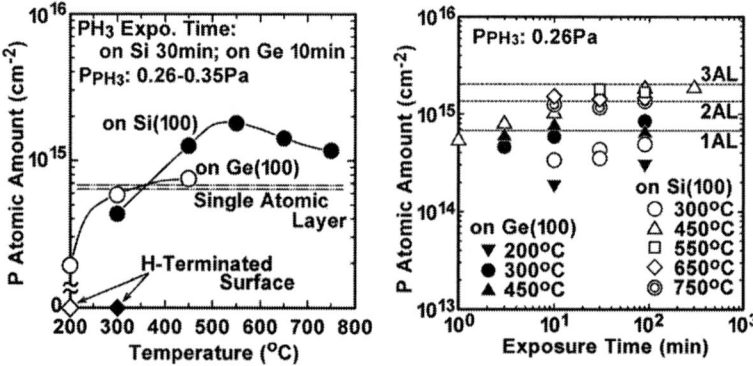

Figure 7 PH$_3$ exposure time and temperature dependence of the P atomic amount on Si(100) and on Ge(100).

saturate below one atomic layer.

It was reported that the saturation P coverage is 1/4 of the number of the Si surface atoms (6.8×10^{14} cm^{-2}) at 300 °C, because the dissociative adsorption of PH$_3$ consumes four surface sites (10). It was also reported that the P coverage becomes lower under high pressure of H$_2$ (11) because of the adsorption of hydrogen even at 350 °C. However, in the present case, the P atomic amount by the PH$_3$ exposure to the hydrogen-free Si surface is about 5×10^{14} cm^{-2}, which is higher than the reported value. It should be noted that the present PH$_3$ partial pressure is three or four orders of magnitude higher than in the report (10).

On the Ge surface at 300-450°C, the P atomic amount tends to saturate to about one atomic layer. On the Si surface at 450-750°C, the P atomic amount tends to saturate to about two or three atomic layers. At 450 °C, the P atomic amount is independent of PH$_3$ partial pressure (0.087-0.78 Pa). Looking at **Fig. 7** in more detail, the P atomic amount is about one atomic layer in early stage.

Si Epitaxial Growth over P Atomic Layer already-formed on Si$_{1-x}$Ge$_x$(100) Surface

In the case of Si growth on the (100) surface with P atomic amount of 2×10^{15} cm^{-2} at a SiH$_4$ partial pressure of 6 Pa at 500 °C (12), surface P atomic amount decreases with increasing SiH$_4$ exposure time without Si deposition. When the P atomic amount decreases below single atomic layer caused by SiH$_4$ reaction, Si growth occurs, and P atoms segregate onto the surface.

At a rather low temperature of 450 °C and a rather high SiH$_4$ partial pressure of 220 Pa, heavy atomic-layer doping of P during Si epitaxial growth was achieved with average P concentration of above 7×10^{20} cm^{-3} and 4 nm thick spacers (12), although the tailing towards surface is also observed and a part of the P atoms segregate or desorb during Si growth. In this case, Si deposition rate may be enhanced by Si$_2$H$_6$ produced from SiH$_4$. Average carrier concentration reaches as high as 4.3×10^{20} cm^{-3} and the resistivity as low as 2.4×10^{-4} Ω-cm. By heat treatment above 550°C, the carrier concentration decreases and the resistivity increases, and they become close to those of the P-doped Si film (13) formed by P diffusion at 1000 °C. It is suggested that the low-resistive P-doped Si film results from higher rate of the Si deposition than that of the electrically inactive P formation such as P clustering. Using this atomic-layer doping technique, a very low contact resistivity of about 5×10^{-8} Ω-cm^2 between Ti and the Si film has been obtained (14). During the growth of the Si cap layer, surface segregation degrades the abruptness of the P doping profile. By lowering the Si growth temperature to 450 °C and increasing the SiH$_4$ partial pressure to 220 Pa, it was speculated that about 1.5×10^{14} cm^{-2} P atoms were incorporated at the initial position (12). By Si cap layer growth using Si$_2$H$_6$ instead of SiH$_4$, 4×10^{14} cm^{-2} P atoms were confined in a Si region of 2 nm thickness (15). But, to achieve abrupt doping profiles at the heterointerface, the P surface segregation has to be effectively suppressed.

Depth profiles of P for the Si/P/Si$_{0.3}$Ge$_{0.7}$ on unstrained Si(100) and 0.8% tensile-strained SOI are shown in **Fig. 8** (16,17). If the initial P atomic amount is below about 2×10^{14} cm^{-2}, it is found that the incorporated P atoms are almost confined within 1 nm around the heterointerface. Moreover, for both the Si/P/Si$_{0.3}$Ge$_{0.7}$ on unstrained Si(100) and on strained SOI(100), the P atom amount is nearly the same as the initial one, in other words, P surface segregation is suppressed. However, in the case of Si/P/Si$_{0.3}$Ge$_{0.7}$ on strained SOI(100), especially with initial P atomic amount higher than 4×10^{14} cm^{-2}, the maximum value of the incorporated P atom amount at around the heterointerface (about

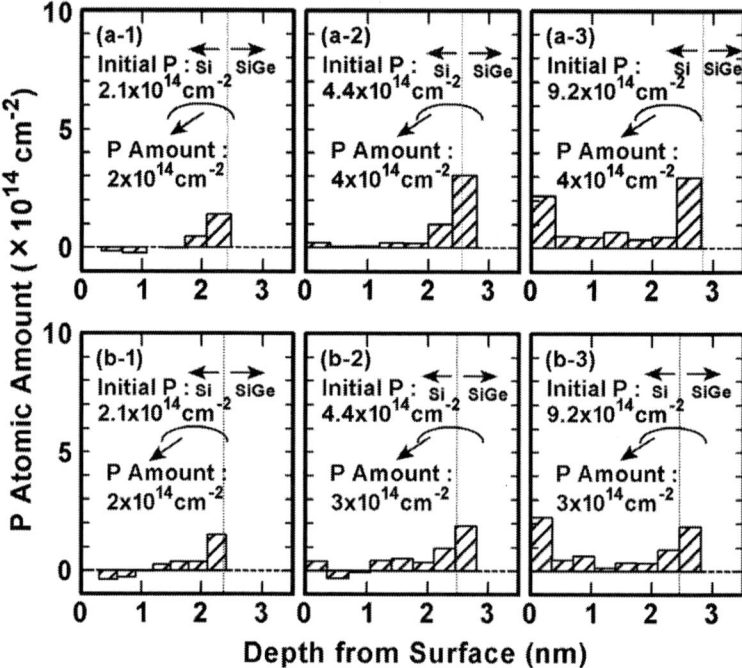

Figure 8 Depth profiles of P for Si/P/Si$_{0.3}$Ge$_{0.7}$ on (a) unstrained Si(100) (upper) and (b) 0.8% tensile-strained SOI (lower). Initial P atomic amount was about (a-1)(b-1) 2.1x10^{14}, (a-2)(b-2) 4.4x10^{14} and (a-3)(b-3) 9.2x10^{14} cm^{-2}. Si cap layer was deposited at 450 $^{\circ}$C at Si$_2$H$_6$ partial pressure of 20 Pa. Interface depths was estimated from decay characteristics of Ge 3d XPS intensity. Because the P atomic amount was calculated by combination of XPS measurement and wet chemical etching of subnanometer-thick Si layer, there are negative values due to experimental error.

3x10^{14} cm^{-2}) is smaller than that of Si/P/Si$_{0.3}$Ge$_{0.7}$ on unstrained Si(100) (about 4x10^{14} cm^{-2}). From these results, it is suggested that tensile strain in the Si substrate enhances P surface segregation and reduces the incorporated P atomic amount at around the heterointerface, although the total P atomic amount on strained SOI(100) is almost the same as that on unstrained Si. In other words, the solid solubility limit of P in Si is reduced by tensile strain.

For a sample of as deposition for the initial P atomic amount about 2x10^{14} cm^{-2}, it is found (**Fig.9**) that the sheet carrier concentration of the Si/P/Si$_{0.3}$Ge$_{0.7}$ on strained SOI(100) is lower than that on unstrained SOI(100), although total incorporated P atomic amount in the Si cap layer on strained SOI is almost the same as that on unstrained SOI . Therefore, it is considered that the strain is impacting the electrical activity of P atoms in Si. Especially in the case on unstrained SOI, the sheet carrier concentration decreases

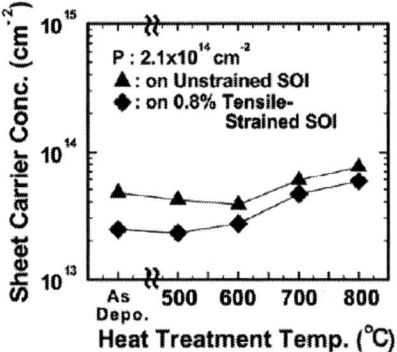

Figure 9 Heat-treatment temperature dependence of the sheet carrier concentration in $Si/P/Si_{0.3}Ge_{0.7}$ on unstrained $Si(100)$ and on 0.8% tensile-strained SOI. Initial P atomic amount was 2.1×10^{14} cm^{-2}.

slightly at $500 \sim 600\ ^{\circ}C$. At $700 \sim 800\ ^{\circ}C$, the sheet carrier concentration increases and the difference becomes smaller. This results from the decrease of maximum P concentration by P diffusion during the heat treatment.

Summary

Atomic layer formation of B or P on $Si_{1-x}Ge_x(100)$ surface and subsequent Si capping layer deposition is achieved at below $500\ ^{\circ}C$.

B doping dose of about 7×10^{14} cm^{-2} is confined within an about 1 nm thick region, but the sheet carrier concentration is as low as 1.7×10^{13} cm^{-2}. The results of in-situ B doping in tensile-strained Si epitaxial growth suggests that the low electrical activity is caused by B clustering as well as the increase of interstitial B atoms.

In unstrained Si cap layer growth on the P atomic layer formed on $Si_{1-x}Ge_x(100)$ with the P atomic amount below about 4×10^{14} cm^{-2} using Si_2H_6 instead of SiH_4, the incorporated P atoms are almost confined within 1 nm around the heterointerface. It is found that tensile-strain in the Si cap layer growth enhances P surface segregation and reduces the incorporated P atomic amount around the heterointerface. The electrical inactive P atoms are generated by tensile-strain in heavy P doped region.

On the other hand, it was confirmed that band engineering for group IV semiconductors becomes possible by strain control of the $Si_{1-x}Ge_x/Si$ heterostructure due to striped patterning (18-20) and that heavy C atomic-layer doping suppresses strain relaxation as well as intermixing between Si and Ge at the $Si_{1-x}Ge_x/Si$ heterointerface (20).

These results open the way for group IV semiconductors with high mobility as well as high carrier concentration by strain control for ULSIs using atomically controlled processing.

Acknowledgements

The study was partially supported by a Grant-in-Aid for Scientific Research from the Ministry of Education, Culture, Sports, Science and Technology of Japan.

References

1. J. Murota and S. Ono, *Jpn. J. Appl. Phys.*, **33**, 2290 (1994).
2. B. Tillack, B. Heinemann, D. Knoll, *Thin Solid Films*, **369**,.189 (2000).
3. J. Murota, M. Sakuraba and B. Tillack, *Jpn. J. Appl. Phys.*, **45**, 6767 (2006)
4. B. Tillack, Y. Yamamoto, D. Bolze, B. Heinemann, H. Rücker, D. Knoll, J. Murota and W. Mehr, *Thin Solid Films*, **508**, 279 (2006).
5. H. Tanno, M. Sakuraba, B. Tillack and J. Murota, *Appl. Surf. Sci.*, **254**, 6086 (2008).
6. H. Tanno, M. Sakuraba, B. Tillack and J. Murota, *Solid-State Electron.*, **53**, 877 (2009).
7. C. W. Ong, H. Huang, B. Zheng, R. W. M. Kwok, Y. Y. Hui and W. M.Lau, *J. Appl. Phys.*, **95**, 3527 (2004).
8. M. Nagato, M. Sakuraba, J. Murota, B. Tillack, Y. Inokuchi, Y. Kunii and H. Kurokawa, 5th Int. SiGe Technology and Device Meeting (ISTDM2010), Stockholm, Sweden, May 24-26, 2010, Abs.No.1918419.
9. Y. Shimamune, M. Sakuraba, T. Matsuura and J. Murota, *Appl. Surf. Sci.* **162-163**, 388(2000).
10. M.L. Yu, D.J. Vitkavage, B.S. Meyerson, *J. Appl. Phys.*, **59**, 4032(1986) .
11. B. Tillack, *Thin Solid Films* **318**, 1 (1998).
12. Y. Shimamune, M. Sakuraba, J. Murota and B. Tillack, *Appl. Phys. Sci.*, **224**, 202(2004).
13. R. B. Fair and J. C. C. Tsai, *J. Electrochem. Soc.*, **124**, 1107 (1977).
14. J. Noh, M. Sakuraba, J. Murota, S. Zaima and Y. Yasuda, *Appl. Surf. Sci.*, **212-213**, 679 (2003).
15. Y. Chiba, M.Sakuraba, J. Murota, *Semicond. Sci. Technol.*, **22**, S118(2007)
16. Y. Chiba, M. Sakuraba, B. Tillack and J. Murota, *Thin Solid Films*, **518**, S231 (2010).
17. Y. Chiba, M. Sakuraba, B. Tillack and J. Murota, 5th Int. SiGe Technology and Device Meeting (ISTDM2010), Stockholm, Sweden, May 24-26, 2010, Abs.No.1918431.
18. J. Uhm, M. Sakuraba and J. Murota, *Thin Solid Films*, **508**, 239 (2006).
19. J. Uhm, M. Sakuraba and J. Murota, *Semicond. Sci. Technol.*, **22**, p.S33 (2007).
20. J. Uhm, M. Sakuraba and J. Murota, *Thin Solid Films*, **517**, 300 (2008).
21. T. Hirano, M. Sakuraba, B. Tillack and J. Murota, *Thin Solid Films,* **518**, S222 (2010).

Physical Modeling of Charge Transport and Degradation in HfO₂ Stacks for Logic Device and Memory Applications

L. Larcher[a], A. Padovani[a], L. Vandelli[a], and G. Bersuker[b]

[a] DISMI, Università di Modena e Reggio Emilia, Reggio Emilia, 42122, Italy
[b] SEMATECH, Albany, New York, USA

The understanding of the physical mechanisms responsible of charge transport and degradation in high-κ stacks is fundamental for the optimization of advanced logic (MOSFETs) and memory (RRAM, DRAM) devices. In this paper, we present a comprehensive physical model describing the charge transport and the degradation/breakdown processes in the HfO₂ layer. This model allows gaining quantitative insights into the physics governing leakage current and degradation processes in HfO₂ stacks, reproducing gate current and TDDB statistics.

1. Introduction

The optimization of high-κ stacks for future CMOS logic device technology requires to reduce the tunneling current and limit the Stress-Induced Leakage Current (SILC) increase, in order to reduce the static power consumption and limit the reliability issues related to the high-κ stack (i.e. aging, degradation, breakdown). Furthermore, the operation principle of resistive switching memories (RRAM), considered a promising technology for next generation Non-Volatile Memory (NVM) devices, is based on the creation of conductive filaments (CF) in the high-κ stacks, which is essentially a controlled dielectric breakdown.

In this scenario, the understanding of the physical mechanisms governing the charge transport and the degradation/breakdown of high-κ stacks is crucial for improving the reliability of advanced CMOS devices and the operation and reliability of RRAM devices. Physics-based models describing charge transport, degradation and breakdown/forming processes in high-κ dielectrics are strongly demanded in order to understand the key factors governing these phenomena, as well as their the temperature-voltage dependencies and statistics.

In this paper we present a comprehensive physical model describing the charge transport and the degradation/breakdown processes in hafnium-based stacks.

The model describing the defect-assisted transport of both electrons and holes in polycrystalline hafnium is described in the Section 2. This model is based on the multi-phonon Trap-Assisted Tunneling (TAT) through defects located preferentially at Grain Boundaries (GBs).

The physical model describing quantitatively the formation of the Hf-rich conductive filament occurring during the forming and breakdown processes in the HfO₂ layer is described in the Section 3. This model, which is based on the multi-phonon TAT conduction mechanism, includes the calculation of the temperature increase due to the microscopic power dissipation at the trap sites and the stress-induced generation of defects responsible of the current increase finally leading to the dielectric breakdown.

Model results are discussed in Section 4. The model allows reproducing accurately the gate current measured across SiO_x/HfO_2 and HfO_2 stacks in static conditions and during the stress experiments leading to the dielectric breakdown. The statistical capabilities of the model allow evaluating the gate current variability and the forming voltage and TDDB distribution.

2. The charge transport model

We model the electron and hole transport across HfO_x stacks by considering that the trap-assisted tunneling through defects preferentially located at GBs is the dominant conduction mechanism. The role played by the GBs in the charge transport has been demonstrated by several independent studies, including scanning tunneling microscopy experiments [1], conductive atomic force microscopy (CAFM) experiments [2], and ab-initio calculations [3]. The model adopted to reproduce the temperature and voltage dependencies of the current is based on the statistical multi-phonon TAT model presented in [4-7].

The gate current accounts for both TAT and direct tunneling (DT) contributions. The hole and electron DT currents from the silicon are calculated by adopting the semi-classical approximation [8], while the electron DT current from the metal gate is computed through the Tsu-Esaki formula [9]. The carrier freeze-out at low temperature is also taken into account [10], while tunneling probabilities are computed through the Wentzel-Kramer-Brillouin (WKB) method.

The TAT current is calculated by summing up contributions of defects, whose position and energy are randomly generated. Such defects can form multiple-trap percolation paths, which are automatically taken into account for the current calculation [12]. The electron and hole current flowing through a percolation path I_p depends on its slowest trap, and is calculated as

$$I_p = \frac{q}{\tau_{c,max} + \tau_{e,max}}. \qquad [1]$$

q is the electron charge; $\tau_{c,max}$ and $\tau_{e,max}$ are the time constants associated to the electron/hole capture and emission by and from the slowest trap of the conductive path, respectively.

The model considers electron and hole coupling to the dielectric phonons [11], which are assumed to be single frequency ω_0 optical phonons [12]. Thus, as sketched in Fig. 1(a), the electron/hole tunneling into a trap is associated with the release of some energy $\Delta E = m\hbar\omega_0$ to the lattice, m being the number of phonons involved. Similarly, the emission of electrons and holes from a trap is associated with the absorption of the energy $\Delta E = n\hbar\omega_0$ from the lattice, n being the number of phonons.

The capture and emission time constants are calculated by accounting for every phonon energy contributions (indexes m and n) [4]:

$$\tau_{c,j}^{-1} = \sum_m N_{j-1}(E_{j,m}) f_{j-1}(E_{j,m}) C_{j,m} P_T(E_{j-1,m}, m) \qquad [2]$$

$$\tau_{e,j}^{-1} = \sum_n N_{j+1}(E_{j,n}) \left(1 - f_{j+1}(E_{j,n})\right) Em_{j,n} P_T(E_{j,n}, n). \qquad [3]$$

E is either the conduction/valence band edge or the energy level of the jth trap; N_j and f_j are the density of states and the Fermi-Dirac occupation probability either at the cathode/anode or at the jth trap; P_T is the tunneling probability; $C_{j,m}$ and $Em_{j,n}$ are the trap capture and emission rates that take into account the carrier-phonon interaction.

$$C_{j,m} = c_0 L(m) \qquad [4]$$

$$E_{j,n} = c_0 L(n) \exp \left(\frac{-n\hbar\omega_0}{kT} \right) \qquad [5]$$

c_0 is a constant depending on the electric field and on the capture cross section of the trap (4); L is the multi-phonon transition probability (5), which models the carrier-phonon coupling.

The physics of the carrier-phonon coupling can be intuitively illustrated and understood using the configuration coordination diagram, which depicts the total energy of the electronic and vibrational states of the system as a function of the generalized lattice coordinate Q, that accounts for the displacements of the lattice atoms induced by charge trapping/detrapping events.

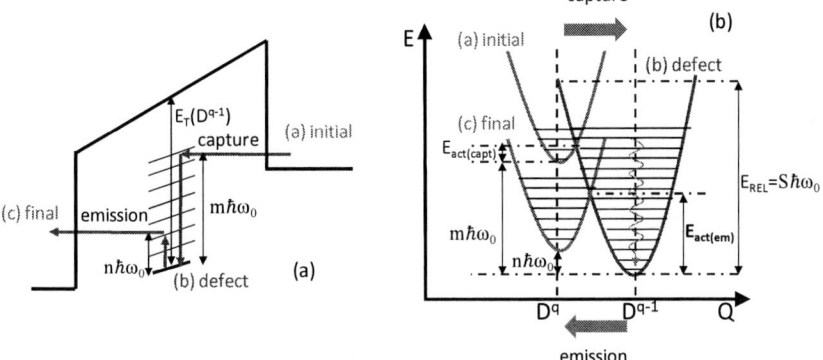

Figure 1. (a) Band diagram illustrating the multi-phonon capture and emission processes occurring during the electron trap-assisted tunneling through a dielectric layer. (b) Configuration coordination diagram depicting the full (i.e. electronic and vibrational) energy of the system in the three phases of the defect-assisted charge transfer.

Figure 1(a) shows the band diagram illustrating the electron trap-assisted tunneling process through a generic dielectric layer. The evolution of total energy of the system including also the vibrational state ones is depicted in Fig. 1(b) for the three phases of the defect assisted charge transfer process: (a) the defect in the initial charge state q, D^q, is empty, and the electron is at the cathode; (b) the tunneling electron is captured by the defect, and the consequent displacements of the lattice atoms is taken into account by the change of the equilibrium configuration $Q(D^q) \rightarrow Q(D^{q-1})$, corresponding to a new charge state q-1 for the defect; (c) the electron is emitted to the anode, and the original lattice configuration is restored. In the whole process, the total energy of the system reduces by the quantity $(m-n)\hbar\omega_0$, which is transferred by the carriers to the dielectric and then dissipated in the form of heat.

For the lattice rearrangements to occur during capture/emission events, the system has to surmount a thermal activation barriers $E_{act(capt)}/E_{act(em)}$, see Fig. 1(b), that sets the temperature dependence of the defect-assisted charge transfer process. $E_{REL}=S\hbar\omega_0$ is the relaxation energy associated with the atomic-scale structural rearrangement required in order to accommodate the carrier charge (13). S is the Huang-Rhys factor, which is an important defect parameter representing the number of the phonons required for such lattice rearrangement (13), thus accounting quantitatively for carrier-phonon coupling.

Defects assisting the electron and hole charge transport are directly linked to the material properties, i.e. to oxygen deficiency inside the HfO$_2$ network. The physical parameters (i.e. density, capture cross section, thermal ionization energy, relaxation energy) are reported in (14) along with effective phonon energy and other material parameters. The thermal ionization and relaxation energies allow identifying the atomistic nature of traps assisting electron/hole transport. Such traps are considered to be randomly distributed in energy and space, and uniform defect distributions are typically considered in simulations.

3. The breakdown/forming model

The model we developed describes quantitatively the physical mechanisms taking place during either the forming operation in HfO$_2$-based RRAM devices and the breakdown process in HfO$_2$ layers. The model describes the formation of a Hf-rich conductive filament inside the hafnium oxide, thus reproducing the exponential current increase observed during electrical stresses (Constant Voltage Stress, CVS, and Ramped Voltage Stress, RVS). The temperature increase due to the local power generation is included through an iterative loop (see Fig. 2) which allows taking into account the generation of new defects related to oxygen vacancy configurations. Thus, the physical characteristics of the conductive filament generated during forming/breakdown are directly linked to the microscopic properties of the HfO$_2$ material.

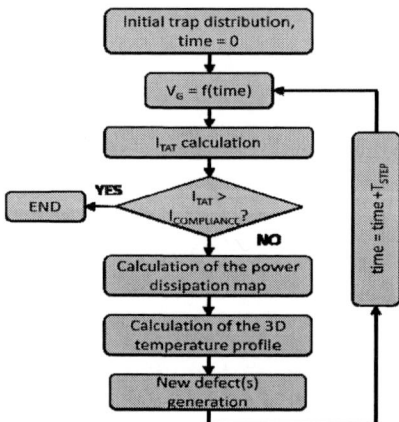

Figure 2. Flow diagram of the model reproducing the evolution of the current flowing through a single GB during the forming operation.

The schematic flow diagram of the model reproducing the current evolution observed during breakdown is shown in Fig. 2.

The model is based on the multi-phonon TAT charge transport model described in the previous section, which is used to calculate the current flowing through positively charged oxygen vacancy defects preferentially located at grain boundaries. The model accounts for the reduction of the defect relaxation energy when the defect density is very high in order to account for the increasingly delocalized nature of the electron wave function, which requires a smaller lattice re-arrangement for the charge accommodation. The simulation ends as soon as the current exceeds the current compliance level. In these conditions we have found that the current is typically driven by a single GB.

The phonon energy released by TAT electrons to the dielectric lattice is calculated for every traps and then it is used to compute the power dissipated inside the HfO_2 layer. Then, the three-dimensional Fourier's heat flow equation is solved to evaluate the temperature profile inside the HfO_2 layer. The heat flow through the surrounding lattice and the metal electrodes is properly taken into account as described in (16).

The presence of both high temperature and electric field at GBs represents the most favorable conditions for the generation of new O vacancy defects, which is included through the effective activation energy description (17)

$$G(T, F) = G_0 \exp\left(-\frac{Ea - bF}{kT}\right).$$ [6]

G(T,F) is the stress-induced defect generation rate that depends on the temperature, T, and the field, F. G_0 is a constant derived from experiments; Ea=4.4eV is the activation energy measured for the defect generation process, which represents the effective energy required for both the Hf-O bond breakage, as theoretically calculated in (17) for the cubic HfO_2; b is the bond polarization factor (17).

Monte Carlo techniques (18) are used to calculate the local generation rates, which depend on the local temperature and field, and new defects are randomly generated in the GB according to these rates. The current, the power dissipation and the temperature are then updated accordingly.

4. Simulation results

Figure 3 and 4 show the gate current densities (J_{gate}) simulated and measured over a wide temperature range (6K - 400K) on two nMOSFET devices with a gate stack comprised of 1.1nm-thick SiO_2 interfacial layer (IL) and either 3nm or 5nm of HfO_2, respectively. As shown, the TAT charge transport model described in Section 2 reproduces accurately the experimental currents in the whole temperature and voltage range. Thus, this model represents a powerful tool to investigate the physical mechanisms governing the charge transport across the stack and its temperature dependence (7), (19).

Interestingly, simulations show that the DT component is negligible (see dashed lines in Fig. 3) in these stacks and the current is mainly due to the electron trap-assisted tunneling mainly assisted by defects located in the IL and HfO_2 film in the thin and thick IL/HfO_2 stack, respectively, see Figs. 3 and 4. The J_{gate} temperature dependence is found to depend strongly on the bias (18). In weak inversion/depletion conditions, it is mainly determined by the carrier supply, described by the Fermi-Dirac occupation probability $f(E)$ in [2] and [3].

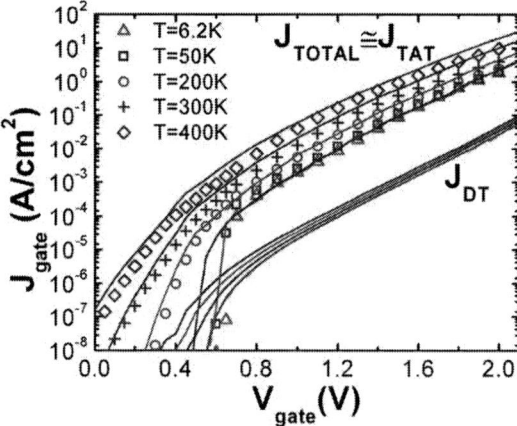

Figure 3. Gate current densities measured (symbols) and simulated (lines) on the nMOS 1.1nm/3nm IL/HfO$_2$ stack at different temperatures. Solid and dashed lines refer to the total current and to the DT component, respectively.

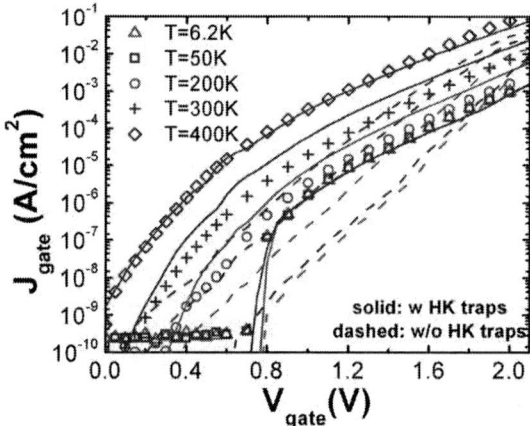

Figure 4. Gate current densities measured (symbols) and simulated (lines) on the nMOS 1.1nm/5nm IL/HfO$_2$ stack at different temperatures. Solid and dashed lines refer to the total current and to the gate current simulations performed neglecting the HfO$_2$ traps, respectively.

Conversely, in strong inversion the temperature dependence is dominated by the electron-phonon interaction, which is accounted for by the multi-phonon transition probability L(m), see equations [4] and [5]. Therefore, reproducing the gate current and its temperature dependency in strong inversion allows deriving important physical parameters related to the atomic configuration of IL and HfO$_2$ defects, i.e. the thermal ionization energy and the relaxation energy (7), (18). The results found from simulations indicate that neutral oxygen vacancies (V^0) are the most probable defects assisting the electron transport in SiO_2, while positively charged oxygen vacancies are the defect configurations assisting the electron TAT in HfO_2 film (7), (18). The same approach can be applied also to pMOS devices to identify the traps assisting the hole transport (18).

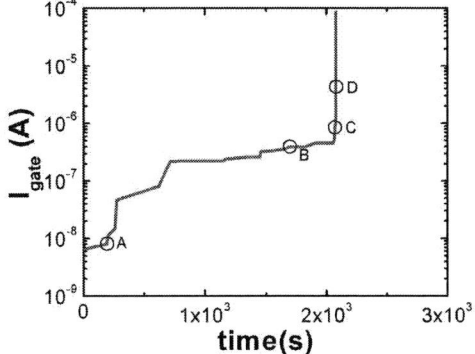

Figure 5. Evolution of the current simulated through a single GB (I_{gate}) in a 7-nm thick TiN/HfO$_2$/TiN capacitor during a Constant Voltage Stress (CVS) performed at $V_G = 2.8V$.

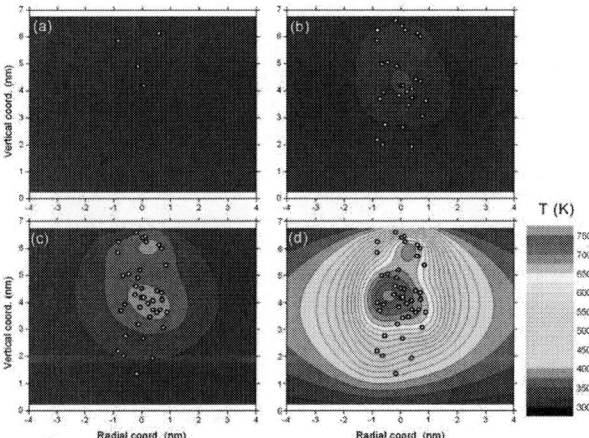

Figure 6. 2D radial maps of temperature in the GB extracted from BD simulation at different phases of the current transient (A, B, C and D in Fig. 5). The maps contains information about the defect locations, represented by the dots.

Fig. 5 shows the simulation of the evolution of the gate current (I_{gate}) driven by a single GB during a CVS ($V_G = 2.8V$) on a 7nm-thick TiN/HfO$_2$/TiN capacitor. Simulations are performed using the described in Section 3. As shown, I_{gate} slowly grows until an abrupt increase corresponding to the breakdown (BD) event is observed. The evolution of the defect generation and temperature increase during the stress is shown in the 2D radial maps in Fig. 6. Each plot corresponds to different phases of the I_{gate} increase transients (labeled as A, B, C and D in Fig. 5). At the beginning of the stress (point A in Fig. 5) only few native defects are present in the GB: the initial TAT current and the related power dissipation are low, and the temperature is approximately uniform and equal to the ambient temperature, see Fig. 6(a). For this reason, the defect generation rate is low and approximately uniform across the entire GB volume, see Fig. 6(b). This situation changes radically when the random generation of additional defects leads to the formation of a favorable percolation path that strongly enhances the local current (point C in Fig. 5) and the related power dissipation. The strong increase in the current raises the local temperature, which in turn enhances the local defect generation rate around the high temperature region, see Fig. 6(c)-(d). Such process triggers a positive feedback that quickly leads to the exponential increase of the current leading to the BD (point D in Fig. 5).

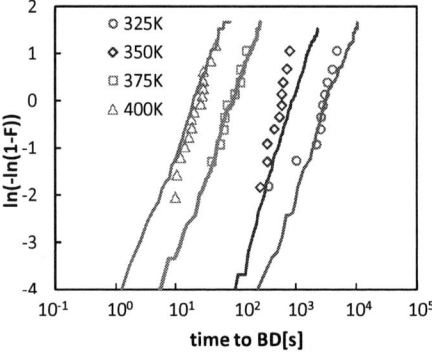

Figure 7. Experimental (symbols) and simulated (lines) time to BD (TDDB) distributions on a 8nm TiN/HfO$_2$/TiN stack at 3.3V at different temperatures.

The statistical capabilities of the model (i.e. defects are randomly generated) allows also simulating the time-dependent dielectric breakdown (TDDB) distributions, thus enabling the prediction of the gate oxide reliability life-time. Figure 7 shows the Weibull plot of the TDDB distributions simulated and measured (20) on a 10^{-5} cm^2 8nm TiN/HfO$_2$/TiN stack at different temperatures. Note that these large area devices contain a large number of GBs. Therefore, BD simulations have been performed by accounting for the current evolution calculated in a large number of isolated, randomly located GBs. The BD is found to occur as soon as the current in a single GB exceeds the pre-defined compliance value (5µA). Again, Figure 7 shows that the model reproduces accurately the temperature dependence of the TDDB distribution, further demonstrating the correctness of the physical models of both electron transport and defect generation. Furthermore, the simulated TDDB distributions follow the Weibull dependency, indicating that the model correctly describes the "weak link" feature, which si a signature of the BD process.

5. Conclusions

In this paper, we have presented a comprehensive physical model reproducing the current flowing across HfO$_2$ stack under static conditions and during the stress-induced degradation experiments. This model allows improving the understanding of the physical mechanisms responsible of charge transport and degradation in HfO$_2$ stacks, which is fundamental for the optimization of advanced logic and innovative memory devices.

Thanks to its statistical capabilities, this model allows reproducing TDDB distribution, thus enabling the life-time prediction of HfO_2 dielectric stack.

References

1. K. S. Yew *et al.*, *Proc. SSDM*, 2009.
2. K. L. Pey *et al.*, *Microelectron. Eng.*, **80**, pp. 353–361, 2005.
3. K. P. McKenna *et al.*, *Appl. Phys. Lett.*, **95**, p. 222111, 2009.
4. L. Larcher, *IEEE Trans. Electron Devices*, **50**(5), pp. 1246-1253, 2003.
5. M. R. Herrmann and A. Schenk, *J. Appl. Phys.*, **77**(9), pp. 4522-4540, 1995.
6. A. Padovani *et al.*, *Proc. IEEE Int. Rel. Phys. Symp.*, pp. 616-620, 2008.
7. L. Vandelli *et al.*, *Proc. of ESSDERC*, pp. 388-391, 2010.
8. N. Yang *et al.*, *IEEE Trans. Elec. Dev.*, **46**(7), pp. 1464-1471, 1999.
9. C. B. Duke, *Tunneling in solids*, Academic Press, 1969.
10. S. M. Sze, *Physics of semiconductor devices*, Wiley-Interscience, 1981.
11. C.H. Henry, D. V. Lang, *Phys. Rev. B*, **15**(2), pp. 989-1016, 1977.
12. M. V. Fischetti, D. A. Neumayer, E. A. Cartier, *J. Appl. Phys.*, **90**(9), pp. 4587-4608, 2001.
13. K. Huang, A. Rhys, *Proc. R. Soc. London*, **204A**, pp. 406-423, 1950.
14. L. Vandelli *et al.*, *IEEE Trans. Elec. Dev.*, in press, 2011.
15. D. Muñoz Ramo *et al.*, *Phys. Rev. B*, **75**, p. 205336, 2007.
16. L. Vandelli *at al.*, to be presented at *Proc. IMW 2011*.
17. J. McPherson *et al.*, *Appl. Phys. Lett.*, **82**(13), pp. 2121-2123, 2003.
18. D. T. Gillespie, *J. Comput. Phys.*, **22**(4), pp. 403–434, 1976.
19. L. Vandelli *et al.*, in press on *IEEE Trans. Elect. Dev.*, 2011.
20. G. Bersuker *et al.*, *IEDM Tech. Dig.*, p. 791, 2008.

198

ECS Transactions, 37 (1) 199-204 (2011)
10.1149/1.3600740 ©The Electrochemical Society

Formation and Characterization of NiYb Silicides

Y. Yuan and E. Ivanov

Tosoh SMD Inc, Grove City, Ohio 43123, USA

We have studied the silicidation of nickel (Ni)-ytterbium (Yb)
films deposited from sputtering target of Ni alloyed with Yb. We
have conducted structural, electrical, and optical characterizations
of the deposited films. Our results indicated the formation and
properties of NiYb silicides were related to the Yb addition to Ni
and the annealing conditions.

Introduction

Recently, new semiconductor materials such as high-K metal gate materials have been
developed and applied to manufacture complementary metal-oxide semiconductor
(CMOS) integrated circuit (IC) devices of 45 nm or less feature sizes with higher
operating speeds and enhanced performance. Such CMOS devices require stable and low
resistivity contact materials. Traditional Co-based and Ti-based silicides are typically
utilized as contacts for 65 nm and above transistor devices. As a consequence of the
unsuitability of Co-based and Ti-based silicides, Ni-based silicides contact technology
has been developed for use in fabricating transistors with feature sizes of 45 nm and
below. However, the contact resistance of NiSi on n-type Si for NMOS is high due to
high Schottky barrier height of ~0.65eV.

Rare earth metals have low electric work function and have been used for low
Schottky barrier applications. Ytterbium (Yb) silicide has been reported to have low work
function and Schottky barrier height [1-2]. It is believed adding Yb to nickel silicide
could reduce NMOS work function to reach NMOS band edge. Pure rare earth metals
(REM) silicides usually form from the solid state interdiffusion of REM and silicon
atoms by furnace or rapid thermal annealing (RTA) the REM-films deposited on silicon
substrate. Ni-REM alloy silicides can be formed from annealing the REM films deposited
on the silicon substrate through either co-sputtering Ni and REM targets or sputtering a
single pre-alloyed Ni-REM target. REM and Ni-REM alloys have shown a great potential
to form stable and low resistance silicides for 45 nm and below transistor devices.

Experimental

Ni-Yb sputtering targets of 99.95% purity and different sizes up to 500mm diameters
have been manufactured at Tosoh SMD, Inc. The Yb contents in NiYb alloy targets vary
from 10at% to 50at%. 30nm thick films were deposited on the silicon wafers from the
Ni17at%Yb sputtering target in a VersaMag/3117 physical vapor deposition (PVD)
system. 5 minute annealing of the Ni-Yb/Si structures at temperatures varied from 250°C
to 750°C has been applied to form the silicides.

The metallic and gaseous trace impurities contents in the precursor sputtering target materials were measured using glow discharge mass spectrometry (GDMS) and LECO

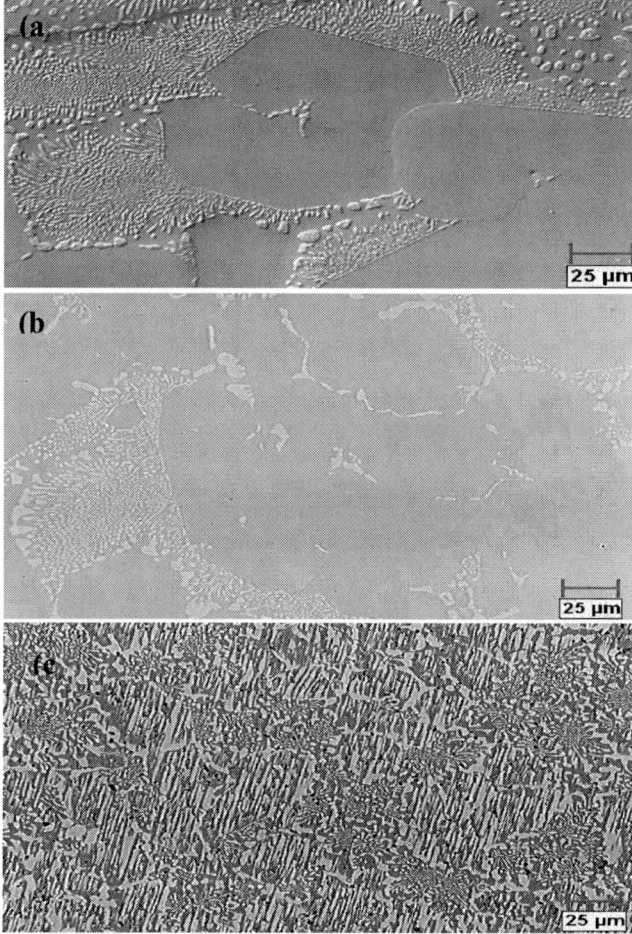

Figure 1. Optical microscopy images of (a) Ni10at%Yb, Ni17at%Yb, and (b) Ni50at%Yb materials. Eutectic structure was seen in the NiYb materials.

methods, respectively. The alloying element contents in sputtering target were measured by inductively coupled plasma (ICP) mass spectrometry. The film resistance was measured using four-point probe. The film reflectance was measured using Varian Cary 300 spectrophotometer. Microstructural examinations were performed using optical microscopy and scanning electron microscopes (SEM). X-ray diffraction (XRD) and energy dispersive spectroscopy (EDS) were performed on the film samples to identify the phase assemblies, compositions, and structures of silicides.

Results and Discussion

Table I shows the trace impurities in the NiYb PVD sputtering target measured by GDMS and LECO methods. The measurement result indicates the sputtering target material has total content of impurities less than 500ppm and meets 99.95% purity standard. REM materials including Yb, Er, and Y are chemically active and show notable reaction with oxygen, moisture, and carbon dioxide gas, so it is usually difficult and challenging to get high purity (>99.9%) REM and alloy materials. Many of the impurities in the precursor target materials can degrade or poison the formed nanostructures inside the transistor devices; therefore it is crucial to get the sputtering target material as pure as possible.

TABLE I. Measured impurities in Ni17at%Yb material.

Element	Concentration (ppm*)	Element	Concentration (ppm)
Si	13.4	Sr	0.054
Al	6.18	Y	10.5
Mo	0.81	Nb	0.002
V	0.26	Ru	1.76
Cr	0.38	Rh	<0.001
Ti	2.05	Pd	<0.006
Ca	13.6	Ag	0.17
Cu	26.6	Cd	0.01
Mg	38.8	In	<0.001
Mn	15	Sn	0.04
Fe	11.6	Sb	0.23
Co	10.7	Te	0.29
Na	0.033	I	0.24
K	<0.005	Cs	<0.001
Li	0.37	Ba	<0.006
Zr	0.33	Hf	<0.04
Be	<0.001	W	1.51
F	<0.009	Re	0.056
P	0.098	Os	0.11
Cl	2.39	Ir	0.065
Sc	<0.001	Pt	<0.005
Zn	2.76	Au	<0.002
Ga	2.52	Hg	<0.003
Ge	24.6	Tl	0.03
As	9.17	Pb	4.55
Se	1.5	Bi	<0.002
Rb	0.018	Th	0.017
H	3.99	U	0.0037
C	42.7	O	36.2
N	8.97	S	<10.0

* Part per million (ppm) weight.

Figures 1 and 2 display the microstructures of NiYb ingot precursor materials. The metallographs, SEM image, and EDS analysis show the NiYb materials are clean and have uniform structures.

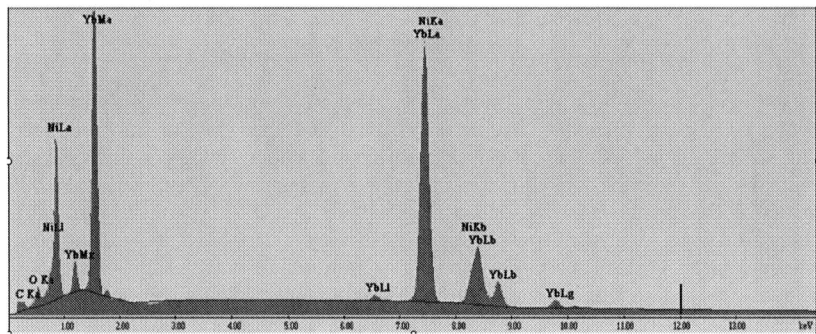

Figure 2. EDS spectrum of Ni50at%Yb material.

Figure 3 shows the resistivity of NiYb films deposited on the <100> Si wafers and annealed at different temperatures. The as-deposited film had a resistivity of 105 μOhm-cm. The resistivity was significantly reduced to below 15 μOhm-cm after the films were

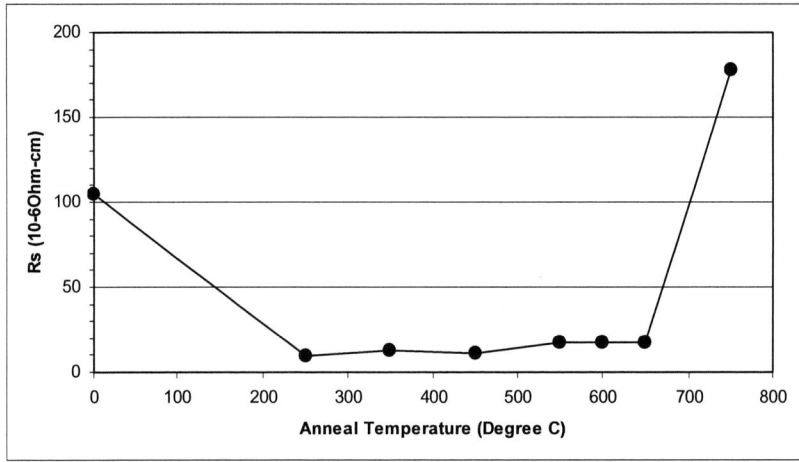

Figure 3. Film resistivity as a function of annealing temperature.

annealed at temperatures below 450°C. The resistivity was slightly increased to about 18 μOhm-cm when the films were annealed in the temperature range of 550°C~650°C. However, there is a dramatic increase in the resistivity for the films annealed at temperatures at 750°C or higher. The film resistivity data indicate the optimal annealing temperature widow for low resistance silicidation is lower than 650°C.

Figure 4 indicates the spectral reflectance curves for the NiYb films. It appears the annealed NiYb films have different reflection behavior compared to the as-deposited NiYb films. Furthermore, the films annealed in the temperature range of 250°C~450°C have similar reflectance curves and so do the films annealed in the temperature of

550°C~650°C. The film annealed 750°C has a reflectance curve similar to bare Si in the band of larger wavelengths.

Figure 5 shows the Bragg Brentano θ/2θ XRD scan pattern for Ni17%Yb sputtering target where the films were deposited from and the Glancing Angle Diffraction (GID) scan patterns for as-deposited and annealed NiYb films. The GID rather than conventional θ/2θ XRD was performed on the films to minimize the penetration of incident x-ray into Si wafer/substrate and enhance the diffractions of NiYb films. The diffraction scans reveal the sputtering target is pure and clean Ni_5Yb intermetallic compound and has well defined crystallographic structure. This intermetallic compound has a theoretical composition of Ni16.7at%Yb, which is consistent with the actual measured target composition of Ni17at%Yb. The x-ray diffraction reveals the as-

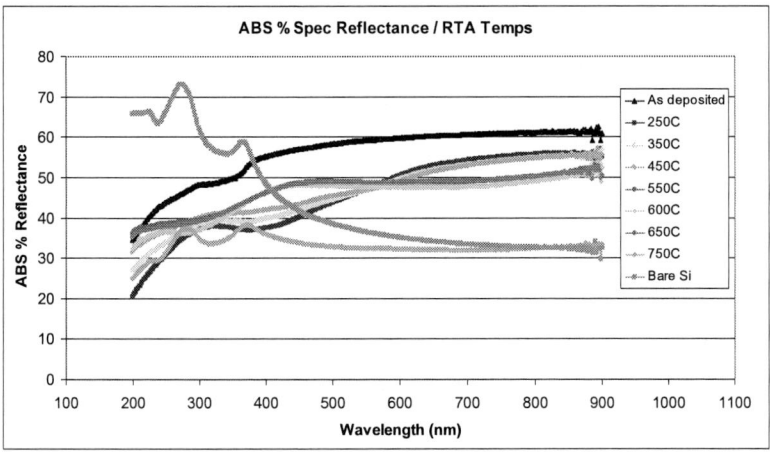

Figure 4. Spectral reflectance curves for Si, as-deposited NiYb film, and annealed NiYb films.

deposited film does not have crystallographic structure and has amorphous or microcrystalline structure instead. YbSi, Yb_5Si_3, and NiSi of low resistivity were formed after the films annealed at 250°C~450°C. YbSi (with different preferred orientations and shift peaks compared to those formed at 250°C~450°C) and high resistance $NiSi_2$ formed when the NiYb films were annealed at 550°C~650°C. The amount of YbSi was significantly reduced when the film was annealed at 750°C. $NiSi_2$ and Ni_2Si were the dominated silicides in the films annealed at this high temperature.

Our results indicated adding Yb to Ni and the formation of Yb_5Si_3 silicides enable the silicide layer to achieve low resistance. In other words, the presence of Yb in Ni and the formation of Yb silicide suppress the formation of high resistance Ni silicide.

The changes in film reflectance may be related to the changes in film surface characteristics including roughness resulting from the phase transformation of Yb and Ni silicides. The NiYb films annealed at 250°C~450°C have the same group of major phases of Yb_5Si_3, YbSi, NiSi, and Ni_2Si. The NiYb films annealed at 550°C~650°C have the same group of major phases of YbSi, NiSi, Ni_2Si, and $NiSi_2$. The major phases in the films annealed at 750°C are $NiSi_2$ and Ni_2Si. The reflectance curves are different for the films annealed at these three different annealing temperature ranges while are very similar for the films annealed at the temperatures within each of the three temperature ranges. The changes in phase assembly can change the film structure and crystallographic orientations and thus change film surface characteristics such as surface height variation or roughness and film reflection behavior.

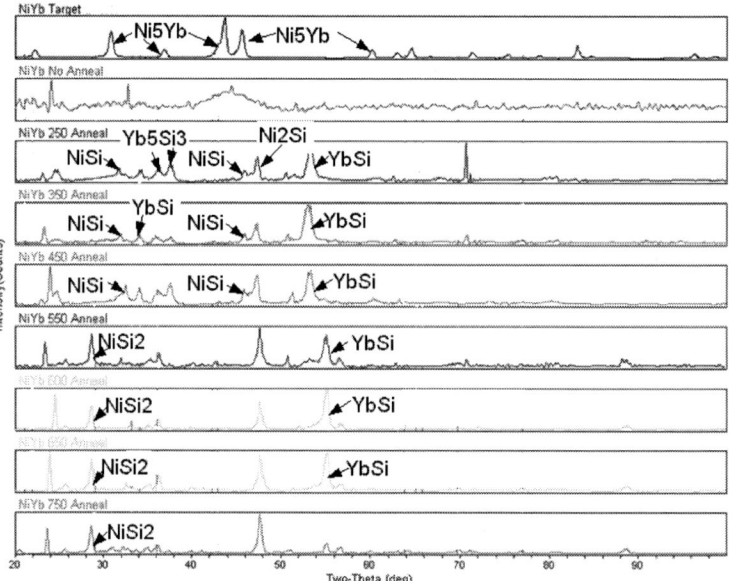

Figure 5. X-ray diffraction scan patterns for sputtering Ni17at%Yb target, un-annealed NiYb film, and annealed NiYb films.

Conclusions

We have developed processes for manufacturing NiYb sputtering targets of different compositions and geometry configurations with target blank sizes up to 500mm diameters. We have investigated the silicidation and material characteristics of the films deposited on <100> silicon. Our results indicate alloying Ni with Yb and forming ytterbium silicides enable the silicide layers to sustain low resistivity. We found different silicide phases formed at different annealing temperatures. Both film resistivity and reflectance were affected by the formation of Yb silicide phases.

References

1. Lee RTP, Lim AE-J, Tan K-M, Liow T-Y, Lo G-Q, Samudra GS, Chi DZ and Yeo Y-C, *IEEE Electron Device Letters* **28**, 2007, pp. 164.
2. S.Y. Zhu, J.D. Chen, M.F. Li et al., *IEEE Electron Device Letters* **25**, 2004, pp. 565.

CHAPTER 6

CHALLENGES IN NEW DEVICES
AND APPLICATIONS

206

Unipolar CMOS Logic for Post-Si ULSI and TFT Technologies

T.P. Ma and Sun Xiao

Yale University, Department of Electrical Engineering, New Haven, CT 06520, USA

This article introduces a novel "unipolar CMOS" logic circuit scheme based on an inverter that consists of two N-channel (or P-channel) enhancement-mode MOSFET's, where one has a positive threshold voltage and the other a negative threshold voltage. In contrast, the conventional CMOS inverter consists of a N-channel and a P-channel MOSFET, where the P-channel MOSFET can significantly limit its overall switching speed, due to the low hole mobility, especially for such III-V semiconductors as InAs, GaInAs, and GaAs, where there is a huge disparity between the electron and the hole mobility. This mobility mismatch issue is even more severe for thin-film transistors that are difficult to make either p-channel or n-channel transistors, such as zinc oxide, amorphous Si, and many organic transistors. The implementation of the "unipolar CMOS" logic will obviate the need for a P-channel MOSFET in the CMOS inverter switch, and takes full advantage of the higher drive current that N-channel MOSFET's offer.

.

Background

The building block for the conventional CMOS (Complementary Metal-Oxide-Semiconductor) logic technology consists of a N-channel MOS Field-Effect Transistor (NMOSFET) and a P-channel MOS Field-Effect Transistor (PMOSFET), as shown schematically in Figs. 1(a) and 1(b), respectively, using SOI (Silicon On Insulator) technology as an example, where S stands for semiconductor. As sketched, these transistors are normally off, and in order to turn on either transistor, a sufficiently large gate voltage needs to be applied on its corresponding gate electrode in reference to its source electrode. The threshold voltage (with respect to its source) that is required to turn on a transistor, Vt, is positive for NMOSFET, and negative for PMOSFET.

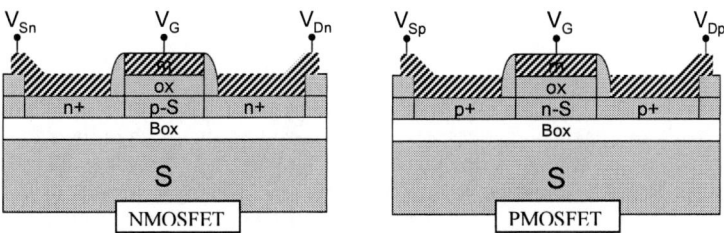

Fig.1 Conventional SOI CMOS Transistors: (a) NMOSFET, (b) PMOSFET

Figure 2 shows the schematic sketch of a logic inverter circuit, made of an NMOSFET and a PMOSFET on a SOI wafer. To illustrate the operating principle, let's assume that the threshold voltage for the NMOSFET (Vtn) is 1.0 V, and that for the PMOSFET (Vtp) is −1.0 V, and the power supply voltage (Vdd) is 3.0 V. Let's also define the logic "1" state (or logic "high" state) to correspond to the voltage range of 2.0 to 3.0 V, and the logic "0" state (or logic "low" state) to correspond to the voltage of 0 to 1.0 V. When the input voltage is 3.0 V (i.e., "high") with respect to ground, the NMOSFET is turned on (because its gate-to-source voltage is 3.0V, which exceeds Vtn), and the PMOSFET is off (because its gate-to-source voltage is 0 V). As a result, Vout is ~0V (i.e., "low"). On the other hand, when the input voltage is 0 V (i.e., "low") with respect to ground, the NMOSFET is turned off (because its gate-to-source voltage is below Vtn), and the PMOSFET is turned on (because its gate-to-source voltage is −3.0 V, which exceeds Vtp). As a result, Vout is ~3.0 V (i.e., "high"). Thus the inverter behaves as expected: a "high" input yields a "low" output, and a "low" input causes a "high" output.

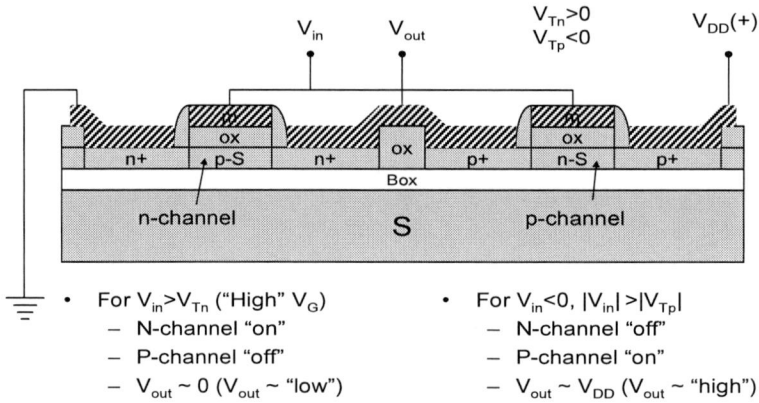

Fig.2 Conventional SOI CMOS Inverter

Note that a key feature of CMOS inverter is that, in either state, the standby power is very small, because one of the transistors is always turned off, and therefore there is little current flowing between the power supply and the ground. And this is the main reason why CMOS technology overtook the NMOS technology for digital applications in the 80's.

The drive current of either the NMOSFET or the PMOSFET is approximately proportional to its carrier mobility and channel width. Since the mobility of electrons (μ_e) is different from the mobility of holes (μ_p) for a given semiconductor, in a CMOS inverter circuit, or a logic circuit based on CMOS inverters, the width of the PMOSFET (Wp) is made differently from that of NMOSFET (Wn), in order that each transistor flows the same amount of current in the CMOS inverter. More specifically, the Wp/Wn ratio is set to be the mobility ratio of μ_e/μ_p for the purpose of current matching. For example, in Si-based logic circuits, the width of the PMOSFET is typically ~2 times that of the NMOSFET, corresponding to the μ_e/μ_p ratio of Si, for current matching.

For a semiconductor that has very large μ_e/μ_p ratio (say a ratio of 20), one must make the Wp/Wn ratio similarly large for the CMOS inverter to realize the current matching condition. In such cases, the large Wp required for current matching results in large PMOSFET transistor size, which increases the chip cost, and makes the overall circuit layout difficult to design. It should be noted that in many III-V semiconductors with high electron mobility, the hole mobility can be 20 times smaller. Similar situation exists in many thin-film transistor and organic semiconductor substrates.

Table 1 shows the electron and hole mobility for several different semiconductors [1,2], indicating large μ_e/μ_p ratios for many of them.

	Si	Ge	GaAs	In$_{0.53}$Ga$_{0.47}$As	InAs
Eg (eV)	1.1	0.66	1.4	0.75	0.35
μ_n (cm^2/v-s)	1,350	3,900	4,600	7,800	40,000
μ_p (cm^2/v-s)	480	1,900	500	350	<500
m*/m$_o$	0.165	0.12	0.067	0.041	0.024

Table 1 Electron and hole mobility for a number of semiconductors (after [1,2]), where InGaAs and InAs exhibit μe/μp ratio > 20.

Key Concept of the Unipolar CMOS Inverter

To overcome the problem caused by the disparity of the electron mobility and the hole mobility, we propose an Unipolar CMOS (U-CMOS) logic concept, which utilizes only NMOSFETs (or PMOSFETs) in a novel CMOS-type of inverter, which can serve as the basis for all logic circuits derived from this inverter, and the bottleneck due to the small hole (or electron) mobility is removed.

The word "Complementary" should be emphasized here, because it is what distinguishes U-CMOS from the conventional NMOS logic where the inverter dissipates full "on" current in one of the two logic states. This high stand-by power was the major problem that killed the conventional NMOS logic technology, which was replaced by the CMOS technology that suffers much less stand-by power.

The U-CMOS utilizes the idea of the double channel capability of a double-gate MOSFET built either with a SOI structure or 3-D vertical channels. For convenience, a SOI version is described here to illustrate the principle of operation of the U-CMOS inverter. As shown in Figs. 3, the conduction channel between the source and the drain for a MOSFET built with a SOI structure, where S denotes a semiconductor, and Box denotes the bottom oxide, could be formed by an inversion layer either at the front dielectric/semiconductor interface, or at the back semiconductor/insulator interface. To turn on the front-channel transistor, the front gate voltage with respect to its source must exceed the threshold voltage of the front channel (Vtf). Similarly, to turn on the back-channel transistor, the back gate voltage with respect to its source must exceed the threshold voltage of the back channel (Vtb).

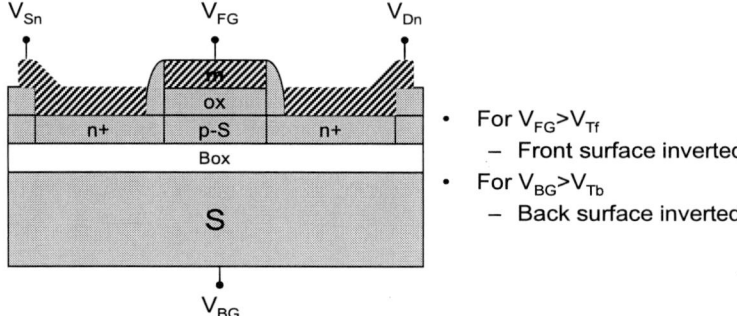

- For $V_{FG} > V_{Tf}$
  - Front surface inverted
- For $V_{BG} > V_{Tb}$
  - Back surface inverted

Fig.3 For a conventional SOI NMOSFET, inversion at either front or back surface of the semiconductor could cause the SOI NMOSFET to turn on.

The transistor in the above example will turn on with either the front-channel transistor or the back-channel transistor inverted. To make the transistor turn on only when the front-channel is inverted, but not when the back-channel is inverted, one can modify the source junction by making it shallower, as shown in Fig.4(a). In this case, when the front channel is inverted. the channel current flows from the drain to the source, just as in the previous device discussed above. However, when the back channel is inverted, the channel current cannot flow from the drain to the source through the back inversion channel, because the n+ region of the source does not reach the back channel, and therefore the source and drain are disconnected. Let's call the transistor shown in Fig. 4(a) the front-channel NMOSFET (or F-NFET). On the other hand, the transistor shown in Fig.4(b) turns on when the back-channel is inverted, as the channel current can flow readily from the drain to source through the back channel, but not when the front-channel is inverted, as the current flowing through the front channel is blocked by the the extended spacer oxide separating the source and the front channel region.. Let's call the transistor shown in Fig.4(b) the back-channel NMOSFET (or B-NFET).

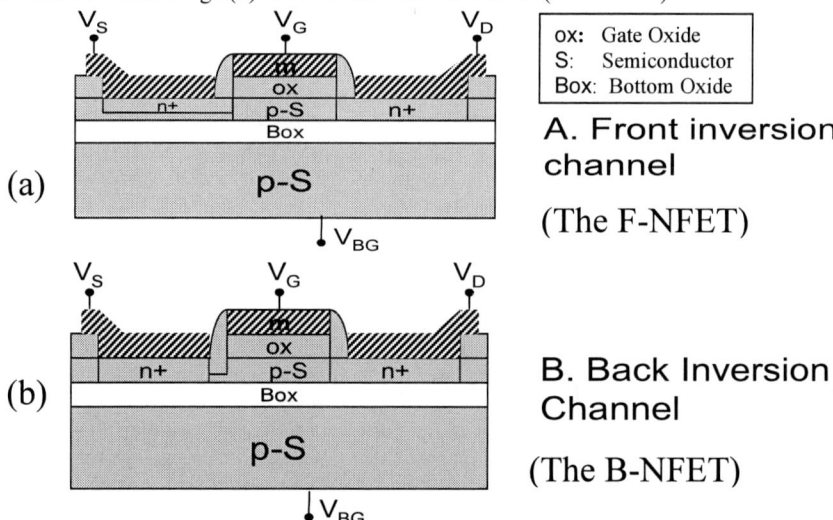

ox: Gate Oxide
S: Semiconductor
Box: Bottom Oxide

A. Front inversion channel

(The F-NFET)

B. Back Inversion Channel

(The B-NFET)

Fig. 4 (a) The F-NFET, is turned on only when the front-channel is inverted; (b) The B-NFET, is turned on only when the back-channel is inverted

Using the above, one can realize an inverter by connecting a F-NFET in series with a B-NFET, very similar to the conventional CMOS inverter, except that the NMOSFET is replaced by a F-NFET, while the PMOSFET is replaced by a B-NFET, plus the presence of an embedded back gate (made of a heavily doped n+ region), as shown in Fig.5.

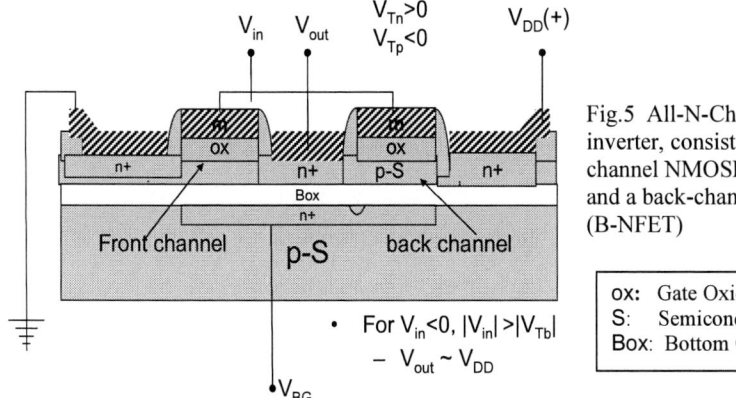

Fig.5 All-N-Channel CMOS inverter, consisting of a front-channel NMOSFET (F-NFET) and a back-channel NMOSFET (B-NFET)

- For $V_{in} < 0$, $|V_{in}| > |V_{Tb}|$
  - $V_{out} \sim V_{DD}$

ox:	Gate Oxide
S:	Semiconductor
Box:	Bottom Oxide

The input voltage is applied to the front gates of both FET's, and its complement is simultaneously applied to the back gate. When the input voltage is "high" (i.e., Vin > Vtf), the back gate is at "low" (i.e. Vbg = 0V), and the F-NFET is turned on while the B-NFET is turned off, which makes Vout "low" (i.e., Vout = 0V). When the input voltage is "low" (i.e., Vin =0 V), the back gate is at "high" (i.e. Vbg > Vtb), and the B-NFET is turned on while the N-NFET is turned off, which makes Vout "high" (i.e., Vout ~ Vdd). Thus the inverter behaves as expected: a "high" input yields a "low" output, and a "low" input causes a "high" output.

From Fig.5, one can see that the cell size of the U-CMOS inverter is further reduced from the conventional CMOS inverter due to the absence of the isolation between p- and n-channels, and the sharing of a common source/drain contact in the middle.

Unipolar CMOS Inverter for III-V Semiconductors

The U-CMOS transistors described above may be modified such that either the front gate insulator, or the back gate insulator, (or both) is replaced by a semi-insulating semiconductor with a larger bandgap than the channel semiconductor. Figure 6(a) shows an example of a front-channel NFET (F-NFET) while Fig. 6(b) shows an example of a back-channel NFET (B-NFET) for III-V semiconductors, where InGaAs has a much smaller bandgap than that of GaP. Figure 7 shows the III-V U-CMOS inverter with only N-channel transistors, which are based on the F-NFET and the B-NFET shown in Fig.6.

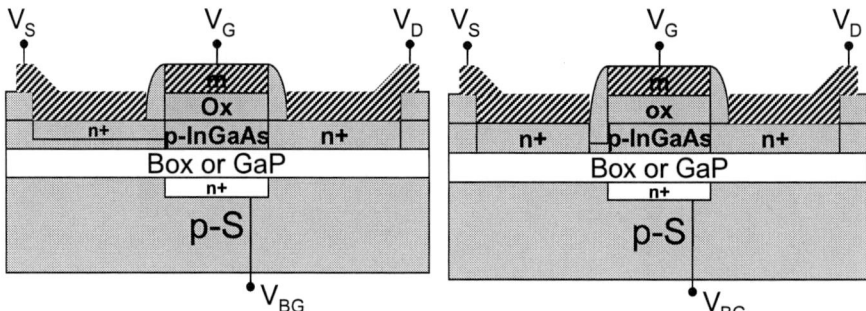

Fig. 6 III-V N-channel NFETs with InGaAs channel and a wide bandgap GaP as a possible bottom insulator: (a) Front-channel NFET (F-NFET); (b) Back-channel NFET (B-NFET). .

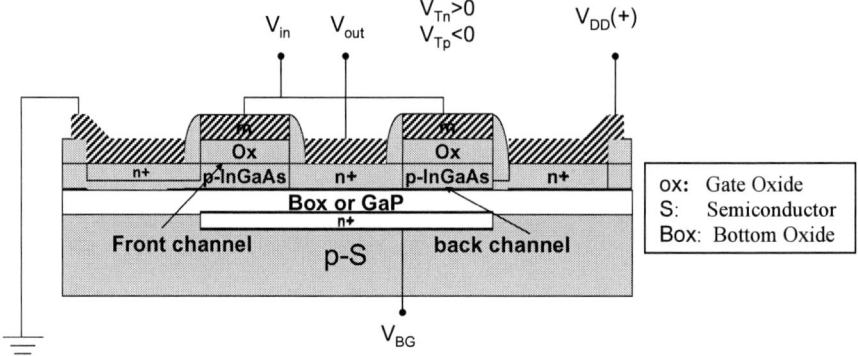

Fig.7 III-V U-CMOS inverter consisting of a F-NFET) and a B-NFET, with InGaAs channel and wide bandgap GaP as possible bottom insulator.

Unipolar CMOS Inverter Implemented in 3-D Double-Gate Structures

The examples shown above are all based on SOI structures for ease of illustration, but they are not necessarily the best designs for many applications, due to the possibility of large parasitic capacitance associated with the back gate. Similar U-CMOS inverter concept may be readily implemented with a 3-D device structure such as the double-gate transistor with a vertical channel, where the left-gate and the right-gate serve the same functions as those of the front-gate and the back-gate, respectively, in the previous examples.

Basic Unipolar CMOS Logic Gates: NAND, AND, NOR, and OR

Based on the unipolar CMOS inverter described above, one can readily construct the corresponding basic logic gates. Figure 8 shows an example of a 2-input NOR gate architecture with complementary inputs. It should be noted that the cell size is relatively small because the Vout contact is shared by 2 transistors, and there is no need for channel isolation. Note that the same NOR gate can also serve as NAND, AND, or OR by Exchanging A,B with ~A,~B, and exchanging GND with Vdd .

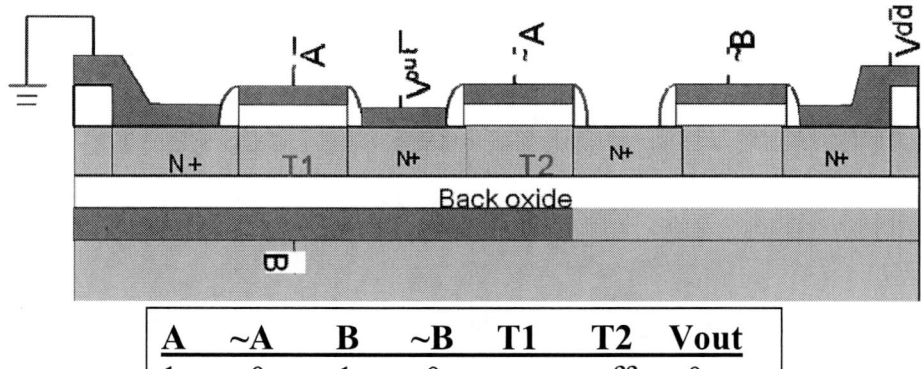

A	~A	B	~B	T1	T2	Vout
1	0	1	0	on	off	0
1/0	0/1	0/1	1/0	on	off	0
0	1	0	1	off	on	1

Fig. 8   Schematic sketch of unipolar CMOS logic NOR gate structure with complementary inputs. The same NOR gate can also serve as NAND, AND, or OR by Exchanging A,B with ~A,~B, and exchanging GND with Vdd .

Unipolar CMOS Logic Gate Implemented in Thin-Film Transistor (TFT) or Organic Transistor Technologies

As mentioned previously, it is very difficult or impossible for many thin-film or organic transistors to find both n-type and p-type channels with sufficiently matched carrier mobilities for conventional CMOS applications [3,4]. Therefore, the possibility of realizing CMOS logic circuits based on the unipolar CMOS concept should be of interest. Figure 9 shows the schematic drawing of a unipolar NAND gate based on the TFT structure with an InGaZnO channel recently published in EDL [5].

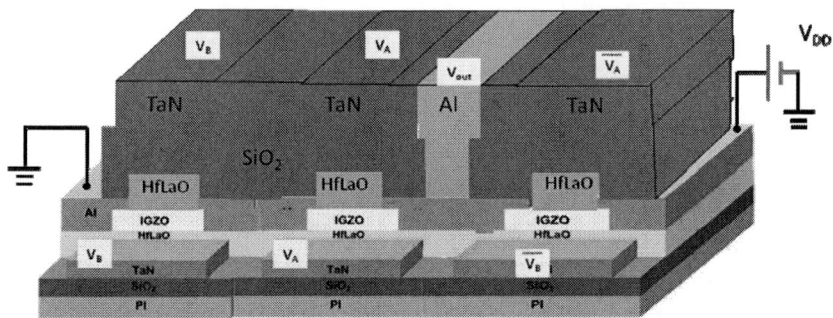

Fig. 9   A schematic sketch of a unipolar logic NAND gate based on the TFT structure with an InGaZnO channel.

Circuit Representation of Basic Logic Gates Based on Unipolar CMOS

Figures 10(a) and (b) depict the circuit representation of the basic logic gates, NAND, AND, NOR, and OR, respectively, based on only n-channel MOSFETs. More sophisticated logic circuits can be built up accordingly.

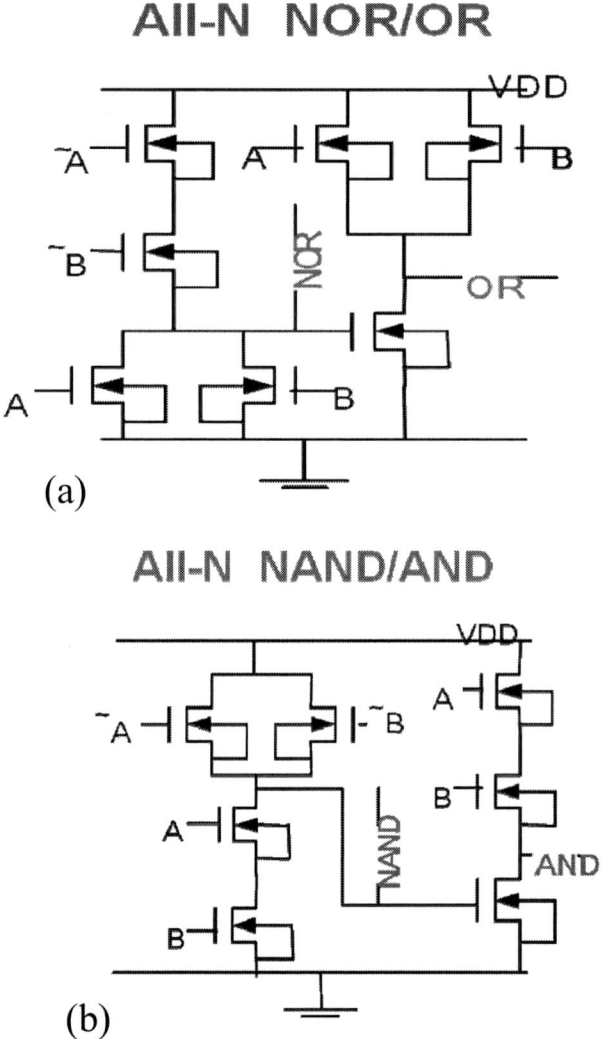

Fig 10   Circuit diagram representation of basic unipolar logic gates based on all n-channel MOSFETs: (a) Two-input all-N NOR and OR gates; (b) Two-input all-N NAND and AND gates

Summary

Many high-electron mobility III-V semiconductors, such as InAs, InGaAs, and InSb, have very poor hole mobility. Similarly, many thin-film transistor channels, including organic semiconductors, lack sufficiently matched electron/hole mobility ratio. In such cases, the logic circuits formed based on the conventional CMOS transistor pair will suffer from layout difficulty, or performance degradation, or both. This article introduces a novel unipolar CMOS logic circuit scheme based on all n-channel (or P-channel) enhancement-mode MOSFET's with double gates, where transistors with both a positive threshold voltage and a negative threshold voltage are provided to eliminate "on" current in standby, just as in the conventional CMOS logic circuits. Examples of unipolar CMOS inverters based on SOI technology have been shown to illustrate the principle of operation, although it makes more sense in practice to implement the unipolar CMOS logic with double-gate transistors built on vertical channel channels. Examples of basic logic gates, such as NAND, AND, NOR, and OR, based on the unipolar CMOS technology have also been presented.

Acknowledgments

This work was partly supported by the National Science Foundation under contract No. MRSEC DMR 0520495.

References

1. Charles Kittel, *Introduction to Solid State Physics*, 7th Edition, John Wiley & Sons, New Jersey (1996).
2. Abigail Lubow, *Study of Inversion Capacitances and Drive Currents for MOSFETs Made of High-Mobility Semiconductors*, Ph.D. Thesis, Yale University, May (2009)
3. Y. Kuo, Editor, *ECS Transactions Thin Film Transistor Technologies 10*, 33(5), Electrochem. Soc., Pennington, 2010.
4. Y. Kuo, Editor, *Amorphous Silicon Thin Film Transistors*, Kluwer Academic Publishers, Norwell, MA, 2004.
5. Nai-Chao Su, Shui-Jinn Wang, Chin-Chuan Huang, Yu-Han Chen, Hao-Yuan Huang, Chen-Kuo Chiang, and Albert Chin, IEEE Electron Device Letters, Vol.31, No.7, July (2010)

Spin-based MOSFET and Its Applications

Y. Saito, T. Inokuchi, M. Ishikawa, H. Sugiyama, T. Marukame, T. Tanamoto

Corporate R&D Center, Toshiba Corporation, 1, Komukai-Toshiba-cho, Saiwai-ku, Kawasaki 212-8582, Japan

We developed a novel spintronic device "Spin-transfer-Torque Switching-MOSFET (*STS*-MOSFET)" which offers non-volatile memory and transistor functions that are CMOS compatible and have high endurance and a fast write time. *STS*-MOSFETs with Heusler alloy ($Co_2FeAl_{0.5}Si_{0.5}$) were prepared and reconfigurability of the novel spintronics-based *STS*-MOSFET was successfully realized in the transport properties. The device showed clear magnetocurrent (MC) and write characteristics with the endurance of over 10^5 cycles. Moreover, the area and speed of million-gate spin FPGAs are numerically benchmarked with CMOS FPGA for 22, 32, and 45 nm technologies including a 20% transistor size variation. We showed that the performance of spin field programmable gate array (FPGA) becomes superior to that of conventional CMOS FPGA as transistor size decreases and MC ratio increases. The overall properties of the *STS*-MOSFETs show great potentialities for future reconfigurable integrated circuits based on CMOS technology.

1. Introduction

The shrinking of silicon (Si)-based conventional complementary metal-oxide-semiconductor (CMOS) transistors will reach its intrinsic physical limits in the future.[1] Since there is a scaling limit of CMOS, new devices capable of adding novel functions to Si-based CMOS transistors should be developed as beyond CMOS devices.

At present memory and logic are two fully separated architectures. Combining nonvolatile storage function and logic functions in a single monolithic device not only solve the current issues such as (1) increase of the power consumption due to the leakage current from the volatile memory and (2) the delay of speed due to the increase of the resistance of global wires generated by transferring information between the memory and logic architectures. Spintronics has attracted much attention as promising candidate to overcome these issues (1) and (2). Spintronics enables remarkable improvement in device performance in terms of nonvolatility, low power consumption, reconstructibility, etc and will offer the possibility of a fully nonvolatile information processing system.

The spin metal-oxide-semiconductor field effect transistors (MOSFET) [2] directly couples the logic element with the nonvolatile memory element, opening up a path to a new kind of ultimate logic-in-memory architecture.

We have proposed a novel spin-based MOSFET "Spin-Transfer-torque-Switching MOSFET (*STS*-MOSFET)" (Fig. 1) [3-5] Combining nonvolatile storage function and logic functions with CMOS compatibility, high endurance and fast write time using spin-transfer-torque-switching (*STS*). Since this structure has magnetic multilayers such as magnetic tunnel junctions (MTJs) or GMR devices on the source or/and drain via tunnel barriers, we can utilize *STS* for the spin manipulation (write process). The critical current for the *STS* is proportional to the area of MTJ (or GMR) device. [6] Therefore small MTJ (or GMR) contribute to the manipulation of spin directions. Moreover, we can utilize the double output signal originating from spin-dependent transport through Si and tunnel magnetoresistance (*MR*) from MTJs, when MTJs are used as the multilayers in Fig. 1. The spin MOSFET directly couples the logic element with the nonvolatile memory element, opening up a path to a new kind of logic-in-memory architecture[7].

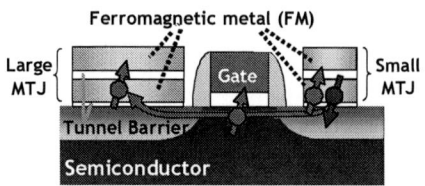

Fig. 1. Spin-based MOSFET of the "spin-transfer-torque-switching MOSFET (*STS*-MOSFET)" type in which MTJs (or GMR) are attached to source and drain electrodes.

However, several issues need to be resolved in order to develop spin MOSFET, such as (1) the lack of good spin manipulation and control method for Si-based spin MOSFET, because the way of spin manipulation proposed by Datta and Das[8] cannot be utilized owing to the small spin-orbit interaction for Si, (2) difficulties of spin injection and detection in the case of a semiconductor such as Si, and (3) preparation of high spin polarized ferromagnet (FM) on Si. In particular, high spin polarization from FM via tunnel barrier should be generated to enhance output signal. In half-metallic ferromagnet, Heusler alloy/MgO would be a promising candidate, because relatively high tunneling spin polarization has already been demonstrated in the cases of granular system and MTJs[9, 10].

The most promising application as a first target for the spin MOSFET such as *STS*-MOSFET would be a reconfigurable logic chip, which can reconfigure the logic data and can be applied to various logic products. Field Programmable Gate Array (FPGA) has a great advantage because a chip is completely programmable and reconfigurable. We have proposed new circuits for the FPGA using *STS*-MOSFET.[11] The area and speed of million-gate spin FPGAs are numerically benchmarked with CMOS FPGA for 22, 32, and 45 nm technologies including a 20% transistor size variation. We have showed that the area is reduced and the speed is increased in spin FPGA owing to the nonvolatile memory function of spin MOSFET. [11] In order to realize the spin FPGA, functions of *STS*-MOSFET should include non-volatile memory and transistor functions with CMOS compatibility such as large on/off ratio of current flow, high endurance and fast write time. The large on/off ratio of current flow of *STS*-MOSFET can utilize when using p/n junction in Si and controlling a gate voltage as shown in Fig. 1.

In these contexts, new innovative ferromagnetic source/drain technologies for next-generation-transistor applications are researched and developed.
First, we demonstrate low interface resistance of Si /tunnel barrier (I)/ ferromagnet (FM) and high spin-polarization of FM. Next, utilizing the Heusler alloy Si/MgO/Co$_2$Fe(Al$_{0.5}$Si$_{0.5}$) (CFAS) interface, we show some data for electrical spin injection and detection in a Si-based device structure. Thirdly, we demonstrate read/write operation and the endurance of *STS*-MOSFET and also present the detailed read and write characteristics of the *STS*-MOSFET. Lastly, we show the result of a benchmark simulation between spin FPGA and CMOS FPGA.
We expect these novel results to lead to technology that opens path for development of next-generation spin transistors.

2. Device fabrication and Experiments

The spin injection devices and back-gate STS-MOSFETs with MTJs were fabricated on Si (001) and p-type Si-on-Insulator (SOI) substrates. We used a heavily doped n-type Si surface (n+-Si) under the tunnel barrier (I)/FM electrodes to reduce interfacial resistance. The regions for n-type channels and n+-Si were formed in the Si substrate by using ion implantation of phosphorus (P) and arsenic (As), respectively, and post-rapid thermal annealing. We are confining spin transport in the active region of the n-type channel. The doping densities N_d in the n-Si channel layer for the spin-injection devices were from ~5×10^{18} to ~5×10^{19} cm^{-3}. For both the spin-injection devices and the STS-MOSFETs, the doping density in the n+-Si layers was 4~5 $\times 10^{20}$ cm^{-3}. A tunnel barrier (MgO, SiO$_x$ or MgO/SiO$_x$) was formed on the Si surface and followed by deposition of a ferromagnetic layer in a UHV sputtering system. The SiO$_x$ layers were formed *in situ* by radical oxidation methods in the chamber. The FM and MgO layers were deposited by DC and RF magnetron sputtering, respectively. The MTJ stack for STS-MOSFETs consists of (Co$_{50}$Fe$_{50}$)$_{80}$B$_{20}$ (CoFeB)/MgO/Co$_2$FeAl$_{0.5}$Si$_{0.5}$ (CFAS)/Ru/Co$_{50}$Fe$_{50}$ (CoFe)/Ir$_{22}$Mn$_{78}$/Ru /Ta. The detail structures of all samples are shown in Table I in Ref. 5. The estimated SiO$_x$ thickness obtained by transmission electron microscopy (TEM) image is 1.0 nm for the condition of the radical plasma oxidization of 50 sec. A large MTJ and a small MTJ were microfabricated on the n+-Si regions by using photolithography, Ar ion milling and reactive-ion etching. The dimensions of the large and small MTJs, used to confirm the STS-MOSFET device operation are 0.35×1.5 μm^2 and 0.3×0.8 μm^2, respectively. Details of the processing for the MTJs are described elsewhere.[12] Current-voltage (I-V) measurements, non-local signal measurements and local MR measurements were carried out for spin-injection devices. Read/write operations and detailed read and write characteristics of the STS-MOSFET were investigated for the back-gate STS-MOSFETs.

3. Junction resistance of FM/I/Si diodes

For highly efficient spin injection and detection into semiconductors, tunneling conduction across the tunnel barriers is very important.[13, 14-19] To obtain spin-dependent transport in a Si channel, interface resistances r_b^* of n+-Si/I/FM should be in a specific range depending on the spin-diffusion length in Si as shown below: [14]

$$r_N \, [\Omega \cdot \mu m^2] \cdot \left(\frac{t_N [nm]}{l_{sf}^{Si} [nm]} \right)^2 \cdot \left(\frac{W[\mu m]}{w[\mu m]} \right) << r_b^* << r_N \, [\Omega \cdot \mu m^2] \cdot \left(\frac{W[\mu m]}{w[\mu m]} \right), \qquad [1]$$

where r_N is the spin resistance of Si channel, t_N is the channel length, l_{sf}^{Si} is the spin-diffusion length in [nm], and W, w is a width of magnetic bar and the Si mesa depth in [μm], respectively. A typical spin-diffusion length in the case of Si is reported experimentally as the $l_{sf}^{Si} \geq 1000$ nm.[3, 16, 20] This value of the spin-diffusion length was estimated in low temperature region (12 K in the case of Ref. 3), however, recently, the long spin-diffusion length of $l_{sf}^{Si} \sim 600$ nm[20] was observed in heavily doped n-type silicon at 300K. On the other hand, from the application point of view, the channel length of the MOSFET in practical use is now less than 32 nm.[1] When taking into these two factors and using the value of spin resistance (r_N) of Si channel as $r_N = 320 \, [\Omega \cdot \mu m^2]$, and assuming W/w~1, the value of the lower limit of Equation [1] is 0.8 $[\Omega \cdot \mu m^2]$. This estimation indicates the r_b^* should be as small as possible and also shows that the issue of conductance mismatch reported in Ref. 14 was solved because of decrease in the channel length of MOSFET. However, from the application point of view, the reduction of contact resistances at source/drain interfaces in MOSFETs also poses a difficult challenge for further scaling small MOSFET.

To evaluate electrical properties of the FM/I/Si contacts, we fabricated two different diodes (Si/I/FM junctions) (junction size: 2×6 μm^2) with and without the n+-Si layer between FM and n-Si. The doping densities are $N_d \sim 5 \times 10^{18}$ cm^{-3} and $N_d \sim 4 \times 10^{20}$ cm^{-3} for n-Si and n+-Si, respectively. The main panel of Fig. 2 shows the absolute value of the current as a function of bias voltage (V) at room temperature for both Diode1 and Diode2. For both diodes, we could not observe the typical rectifying behavior of a conventional Schottky diode, whereas we find almost symmetric behavior with respect to V polarity. The almost symmetric behavior in I-V curves represents the metallic nature in both Si/I/FM junctions; therefore, $N_d \sim 5 \times 10^{18}$ cm^{-3} is enough for decreasing the width of Schottky component of depletion layer in Si. We note that the reverse-bias ($V < 0$) current for n-Si/I/FM junction is extremely small less than two orders of magnitude compared with that for n+-Si/I/FM junction. This is because the depletion width of n+-Si becomes thinner and the conduction band edge of n+-Si moves toward lower energy level as shown schematically in the inset of Fig. 2. The result obtained indicates the tunnel conductance was strongly dependent on doping level of Si surface, showing low resistances for the junctions on n+-Si. Figure 3 shows the summary of the resistance area products (RA) as a function of N_d with various Si/I/FM diode structures. The details of the diode structures are shown in Table I in Ref. 5. The applied bias voltage was $V = -100$ mV. We obtained relatively low RA of 36 $\Omega\mu m^2$ for the junctions on the n+-Si layer, whereas the RA value for those on the lightly doped Si ($N_d \sim 5 \times 10^{18}$ cm^{-3}) was 38k $\Omega\mu m^2$. These results indicate that the RA values of Si/I/FM are strongly dependent on N_d owing to change of the width of a depletion layer of Si. We also confirmed a dependence of the resistance on the SiO$_x$ tunnel barrier thickness in the case of high N_d,[5] showing successful control of the Si/I/FM diode junction resistance.

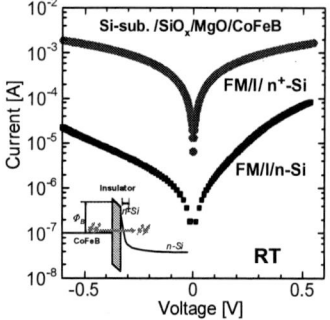
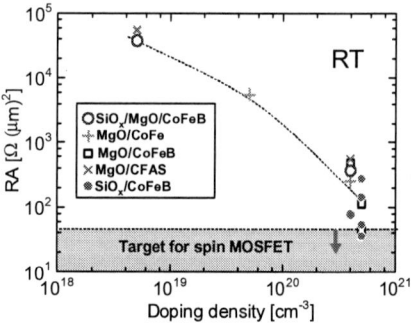

Fig. 2. I-V characteristics of the fabricated Si/SiO$_x$/MgO/CoFeB diodes at 300 K.

Fig. 3. Resistance-area products (RA) vs. doping density in Si(100) with varying insulator (I) barrier/ferromagnet (FM) electrodes at 300 K.

4. highly spin-polarized Co$_2$Fe(Al$_{0.5}$Si$_{0.5}$) and spin-dependent transport through Si

Full-Heusler alloys, CFAS, feature theoretically predicted high spin polarizations for both the ordered L2$_1$ structure and the disordered B2 one (Ref. 21, 22 and reference in). Figure 4 shows summary of the tunnel MR (TMR) ratio on Si substrates as a function of RA value for Si-sub./Ta/(Co$_{50}$Fe$_{50}$)$_{80}$B$_{20}$ (CoFeB)/MgO/(upper FM layer)/Ru/Co$_{50}$Fe$_{50}$ (CoFe)/Ir$_{22}$Mn$_{78}$. Fabricated MgO tunnel barriers on amorphous CoFeB were (100)-

textured and the CFAS films on MgO tunnel barrier had a polycrystalline B2 structure for the post-annealing at 360°C, which was confirmed by cross sectional TEM and nano-beam electron diffraction. The maximum TMR ratio was 220% when the upper FM layer was Fe (0.4nm)/CFAS. Half-metals can be considered to be the ideal materials for spintronics, since the efficiency of spin-dependent devices will be greatest if the current is 100% spin polarized.[23] However, the strong temperature dependence of spin polarization of all reported half-metallic materials, mainly due to spin wave excitation[24-26] and the narrow energy separation between the Fermi level and the conduction or valence band edge,[27, 28] limits their practical applications. Sakuraba et al.[29] reported a series of exciting results concerning Co_2MnSi (CMS) and proved its half-metallic character at a low temperature. Unfortunately, the spin polarization of CMS also sharply decreases with increasing temperature resulting in a strong temperature dependence of *TMR* ratio for MTJs with CMS electrodes. The recent theoretical study on the stability of magnetic structures in the surface and interfaces of Heusler alloys, Co_2MnAl and Co_2MnSi shows that instability of Co moments at the interface is responsible for the strong dependences of the magnetoresistance ratio on both the temperature and bias voltage, which was measured in the MTJs of $Co_2MnAl/MgO/Co_2MnAl$ and $Co_2MnSi/MgO/Co_2MnSi$.[30] This interface instability would disappear, if a ferromagnetic transition metal such as Fe, Co, FeCo were inserted between Heusler alloys and MgO tunnel barrier. This idea is our motivation for the insertion of ferromagnetic transition metals between Heusler alloy (CFAS) and MgO as shown in Fig. 4. However, in the case of CFAS, insertion of thin magnetic layer had little effect on the TMR ratio as shown in Fig. 5. Recently, experimental study[31] revealed that CFAS exhibits the highest effective spin polarization at 300 K and the weakest temperature dependence of spin polarization among all known half-metals. Further study has shown that spin polarization of CFAS decays with increasing temperature (T) following $T^{3/2}$ law perfectly, which indicates that the depolarization of CFAS is determined by spin-wave excitation only. The fact that insertion of thin magnetic layer had little affect on the TMR ratio in CFAS is thought to be related to the nature of T-dependence of CFAS determined by spin-wave excitation only.[31]

Figure 5 shows the thickness of MgO-buffered layer dependence on TMR ratio for MgO-bufer/$Co_2FeAl_{0.5}Si_{0.5}$ (CFAS)/MgO/CFAS/$Ir_{22}Mn_{78}$/Ru MTJs. After deposition of bottom CFAS, we carried out *in situ* annealing at 360°C and after deposition of all stacks of the films, we also carried out the post-annealing at 360°C. The CFAS/MgO/CFAS MTJs on an MgO-buffered Si substrate also show relatively high TMR ratios up to about 120%; however, the TMR ratios strongly depend on the buffered MgO layer thickness on Si substrate as shown in Fig. 5. In the case of thin buffered MgO layer (MgO buffer (<1 nm)/CFAS/MgO (1 nm)/CFAS MTJs), in which MOS operation with large on/off ratio of current flow can be observable, TMR ratio was less than 40% as shown in Fig. 5. Therefore, for the demonstration of *STS*-MOFET, we chose to use the $(Co_{50}Fe_{50})_{80}B_{20}$ (CoFeB)/MgO/$Co_2FeAl_{0.5}Si_{0.5}$ (CFAS)/Ru/$Co_{50}Fe_{50}$ (CoFe)/$Ir_{22}Mn_{78}$ structure on a SiO_x-buffered Si substrate, because we observed relatively large TMR ratio (TMR=62%) even for the thin tunnel barrier of MgO in MTJs with the low RA value of RA=9.6 $[\Omega \cdot \mu m^2]$. The fabricated MgO tunnel barriers on amorphous Si sub./SiO_x tunnel barrier/CoFeB were (100)-textured and the CFAS films on MgO tunnel barrier had a polycrystalline B2 structure. When utilizing the coherent tunneling of MgO tunnel barrier using the MgO buffer on Si substrate and Half-metallic nature of CFAS, great care should be taken in preparing the (100)-textured MgO buffer layer on Si substrate.

Next, we show the electrical spin injection and detection using n-Si/n+-Si/MgO/CFAS

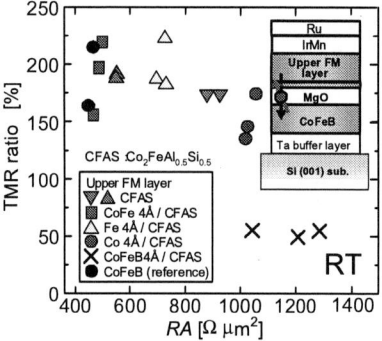 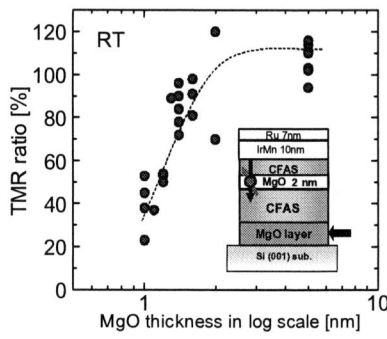

Fig. 4. TMR ratios vs. *RA* values with varying upper electrodes for the polycrystalline Bottom-CoFeB/MgO/Top-CFAS MTJs on Si substrate with and without interfacial insertion layer between Top-CFAS and MgO barrier.

Fig. 5. TMR ratio dependence on the thickness of buffered MgO Si(001) substrates. Inset of figure shows the CFAS-MTJ structure on MgO-buffered Si(001) substrates.

tunnel contacts. As shown in inset of Fig. 6 (a), we fabricated four-probe lateral devices. The structure of the film we used here is that of diode9 (MgO0.6nm / CFAS6nm / Ru24nm). The Si mesa width of the lateral device determined by the etching or/and the implantation of the dopant ions is 1 μm. By changing the width of MgO/CFAS bars (0.4 μm and 0.8 μm), we could distinguish between magnetization reversal processes of the spin injector and detector in nonlocal and local voltage measurements. Here, as shown in Fig. 6, we focus on the local measurements we observed for the current flow shown in Fig. 7(a), which is the same as the current flow of spin MOSFET operation. The measurements were performed by a dc method for the current-voltage scheme, where external magnetic fields (H) were applied parallel to the long axis of the contacts in the film plane. Figure 6(b) and inset of Fig. 6(b) show a representative $MR = \Delta R/R_{minimum}$ as a function of bias voltage at 12 K, and MR curves, respectively, where ΔR and $R_{minimum}$ are resistance difference and minimum of observed resistance (antiparallel magnetization state in the inset of Fig. 6(b)), respectively. MR ratio of about 10% at 12 K was observed in Si channel length of $L = 10$ μm. As shown in Fig. 6(b), the MR ratio was not sensitive to the bias voltage. We could not observe the signal for the current flows between FM1 and Ohm1, and between FM2 and Ohm2, and therefore, we believe that the signal observed in Fig.6 is the spin-dependent signal through Si. As shown in Ref. 3, we could observe the nonlocal signal correspond to the local MR signals. Here, we should discuss the comparison between interface resistance at which we observed local MR signals and the theoretical predicted specific range at which local MR signals should be observed. When using the spin resistance (r_N) for Si channel in Eq. (1) as $r_N = 26000$ [$\Omega \cdot \mu m^2$], which is the resistance of P-doped n-Si for the concentration of $10^{17} cm^{-3}$, the estimated specific range of the interface resistance r_b^* of CFAS/MgO/n+-Si should be 7.8 [$\Omega \cdot \mu m^2$] $\ll r_b^* \ll 7.8*10^4$ [$\Omega \cdot \mu m^2$] in the case of $t_N = 10$ μm and W/w=3 (in the case of the diode9 sample of Figure 6(a) and Table I). As shown in Fig.3, the interface resistance r_b^* [14] of n+-Si/MgO/CFAS for diode 9 structure is 5×10^2 [$\Omega \cdot \mu m^2$]. The value of 5×10^2 [$\Omega \cdot \mu m^2$] is consistent with the estimated specific range of the interface resistance r_b^*. However, we must confirm the corresponding Hanle signals in order to increase confidence that the local signals and nonlocal signals are spin-dependent signals through Si channels. In terms of temperature dependence, unfortunately, the signal became quite noisy at around 150 K. To get a spin device with high-temperature operation, we should further explore not only highly efficient spin injection and detection across the n-Si/n+-Si/FM but also optimizations of the doping technique, device fabrication processes, and the device geometry. An observation of room-temperature spin transport in Si for the local geometry as shown in Fig. 6(a) will be required, which can open a path to Si-based spintronic devices such as spin-MOSFETs.

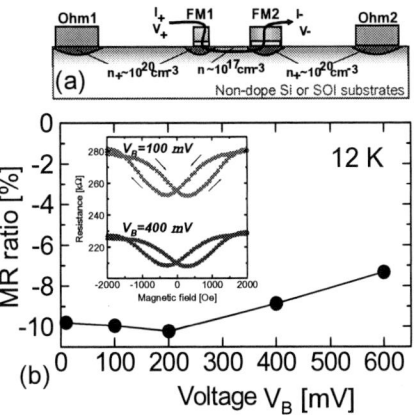

Fig. 6. The local MR ratio as a function of bias voltage for lateral CFAS/MgO/Si channel/MgO/CFAS samples at 12 K. The MR ratio of ~10% was observed. Inset shows the MR curves for V_B=100 mV and 400 mV.

5. Demonstration of read/write operation and detailed read/write characteristics

A bottom-gate MOSFET with MTJs (*STS*-MOSFET) was fabricated by using a p-type (001) silicon-on-insulator (*SOI*) substrate as sketched in Fig. 7(a). Here, the buried oxide (*BOX*) layer and the bottom Si substrate of the *SOI* substrate were used as a gate insulator and a gate electrode, respectively. The detailed MTJ structure of *STS*-MOSFET is ($Co_{50}Fe_{50})_{80}B_{20}$ (CoFeB)/MgO/$Co_2FeAl_{0.5}Si_{0.5}$ (CFAS)/Ru/$Co_{50}Fe_{50}$ (CoFe)/$Ir_{22}Mn_{78}$/Ru /Ta. The thicknesses of all layers of MTJ are designed to switch by *STS* and the synthetic pinned layer structure was employed for cancellation of stray field form the pinned layer. Typical value of magnetoresistance (*MR*) and *RA* for H_{hard} =0 Oe in MTJs are *MR*=62% and *RA*=9.6 [$\Omega \cdot \mu m^2$]. The SiO_x tunnel barrier prepared by the radical oxygen process between n+-Si and MTJ was employed because we obtained the low *RA* value as shown in Fig. 3. The size of Si mesa width (W) was designed to W = 600 μm in order to get a sufficiently large value of the current flow, which is necessary for the write operation of the *STS*. Figure 7(b) shows a cross-sectional transmission electron microscopy (TEM) image of the fabricated *STS*-MOSFET. From this point onward, we show the data measured at room temperature, and therefore, observe *MR* due to spin-dependent local transport only originating in the tunnel *MR* from MTJs on source/drain. The drain current measured between the large MTJ and the small MTJ showed clear change corresponding to the magnetization direction change of each ferromagnetic layer. Figure 8 (a) shows the magnetocurrent (*MC*) ratio as a function of drain voltage (V_D). *MC* ratio was defined as *MC* ratio = ($I_D^P - I_D^{AP}$) / I_D^{AP})×100, where I_D^P and I_D^{AP} are the drain current corresponding to the parallel and antiparallel configurations of each ferromagnetic layer, respectively. The *MC* ratio was about 3% at low V_D. *MR* ratio and *RA* value from MTJ were 62% and 9.6 [$\Omega \cdot \mu m^2$], respectively, as described before. The small *MC* ratio (= (R^{AP}-R^P)/R^P×100 =MR ratio) is attributable to the resistance of parallel magnetization configuration (R^P) is mainly determined by the resistance of Si channel, that is, resistance of MTJ was designed to be a small value (RA=9.6 [$\Omega\cdot\mu m^2$]) in order to realize high-speed operation and to utilize *STS* for write operation. As shown in Fig. 8(a), when the small MTJ was connected as a source, the *MC* ratio became large compared with the structure in which small MTJ was connected as a drain. The large *MC* ratio is in accordance with the operation principle of pseudo-spin MOSFET proposed by Sugahara.[32] The I_D is proportional to the terms;[33] $I_D \propto (V_g - V_S - V_{th})^{\alpha}$ ($1 < \alpha < 2$), where V_g, V_S, V_{th} is gate voltage, the voltage applied to the MTJ, the threshold gate voltage at which I_D start to flow, respectively.

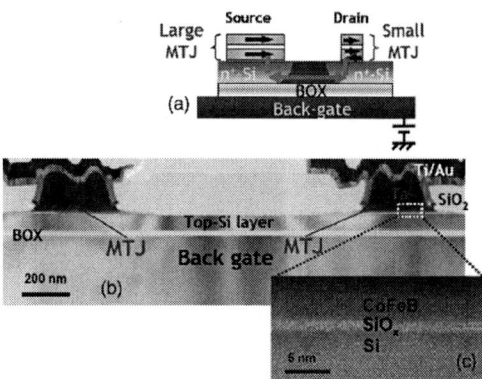

Fig. 7. (a) Illustration of back-gate spin-based MOSFET with MTJs (*STS*-MOSFET) and (b) cross-sectional TEM image of back-gate *STS*-MOSFET. Two MTJs are used for the source/drain. Inset of photograph shows the high-resolution TEM images for the interface of Si substrate/SiO$_x$ on the CoFeB/MgO/CFAS-MTJ.

When small MTJ is connected as a source, relative gate voltage ($V_g - V_S$) changes depending on the resistance of MTJ as shown in Fig. 8(b). When the MTJ is connected as a source, the effective gate voltage becomes low because the voltage drop occurs at the current path in the MTJ, so that the drain current decreases as shown in Fig. 9(a). On the other hand, the *MC* ratio increases because the change of the effective gate voltage, which originates from the *MR* ratio in the MTJ, amplifies the change of the drain current as shown in Fig. 9(b). In the case that the drain current is measured between the large MTJ and the small-MTJ, the resistance of the small MTJ is higher than that of large MTJ, so that the change of the effective gate voltage becomes larger and the *MC* ratio increases when the small MTJ is connected as a source. It should be noted that ideal spin MOSFET has large *MC* ration and large I_D, where large I_D indicates the high-speed operation; therefore great efforts should be made to achieve larger *MC* ratios.

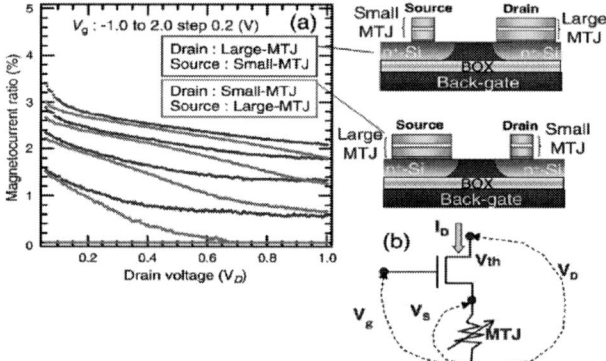

Fig. 8. (a) Dependence of *MC* ratio on V_D measured between large MTJ and small MTJ and (b) the schematic circuit diagram for explaining the origin of the enhancement of *MC* ratio when the small MTJ was connected as source.

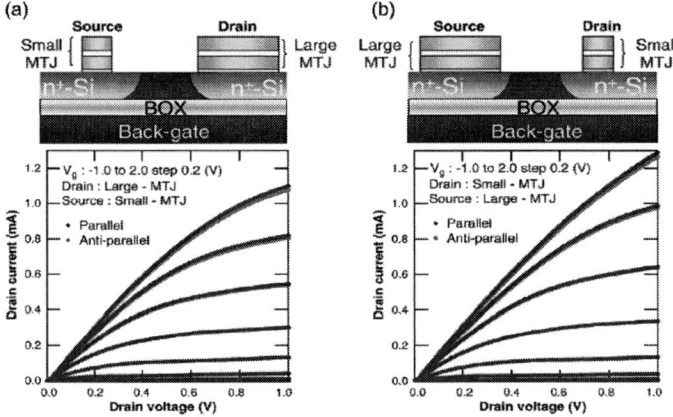

Fig. 9. I_D-V_D characteristic measured between large MTJ and small MTJ (a) small MTJ was connected as drain (b) small MTJ was connected as source.

operations because the hard-axis field reduces the critical current density required for STS.[12] Figure 10 shows the typical switching behavior between the antiparallel and parallel states. The clear switching between the antiparallel states and the parallel states by the STS was observed in the conditions of V_g = 2 V, H_{hard} = 90.0 Oe, H_{easy} = -6.0 Oe. The pulse duration used in this measurement was 100 μsec. The decrease in the value of J_c with increasing H_{hard} is observed. The estimated critical switching current density (J_{c0}) and energy barrier ($E^{ave.}/2k_BT$) of the STS from the pulse duration dependence on J_c in the case of H_{hard} = 90.0 Oe was J_{c0} = 1.2×10^6 A/cm^2 and $E^{ave.}/2k_BT$ = 29. Figure 10 shows read and write endurance of over 10^5 cycles without the breakdown of the tunnel barriers. The stable STS effect of the MTJ occurred in the MOSFET under V_g = 2 V, H_{hard} = 90.0 Oe, H_{easy} = -6.0 Oe. When applying H_{hard}, we can observe stable switching reproducibility as shown in Fig. 10. Thus, we first confirmed good reproducibility of write operation, i.e., resistance changes by STS for the STS-MOSFET that has the MTJs directly on the n+-source/drain.

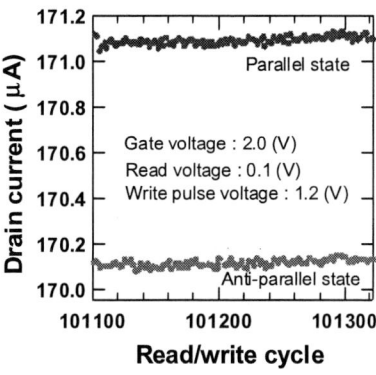

Fig. 10. Resistance changes by STS during >100,000 cycle test in the case that the read and write pulse voltages of 0.1 V and ±1.2 V are applied under V_g = 2.0 V.

6. Benchmark simulation of spin FPGA

6-1. Spin FPGA

FPGA has a great advantage due to the fact that a chip is completely programmable and reconfigurable. However, a conventional FPGA includes a large amount of static random access memory (SRAM), which is a volatile memory composed of six transistors and demonstrates the fabrication limitation of a Si MOSFET. Thus, as one of applications for spin MOSFET, a new FPGA based on novel devices has been expected. SRAM (six transistors) can be replaced by one spin MOSFET. Many SRAMs are used in FPGA such as in lookup tables (LUTs) and the interconnect area of pass transistors. Therefore, this replacement reduces the number of transistors and the FPGA area. Because the speed of FPGA is governed by the length of the wire part, a smaller area of spin FPGA leads to a faster performance.

Figure 11 shows our spin LUT structure[11, 34] for 4-inputs and 1-output, which is a typical set of LUT parameters.[35] The transistor sizes of amplifiers are adjusted such that the input pulse signal is appropriately transferred to the output pulse signal of LUT. We also propose a new spin control pass transistor depicted in Fig. 12. SPICE simulations show that the speed of the pass transistor in Fig. 12 is on the same order as that in the case of ordinal pass transistor connected be the SRAMs.

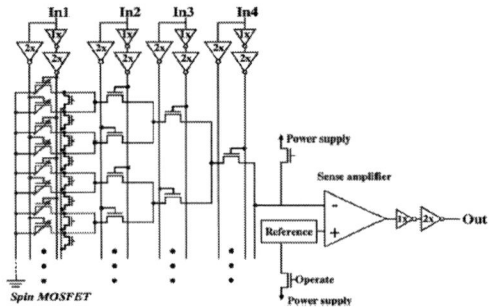

Fig. 11. Schematic of a 4-input lookup table based on the spin MOSFET (spin LUT). Spin MOSFETs replaces SRAMs at the left most part on this figure.

Fig. 12. New routing pass transistor using spin MOSFET.

6-2. Benchmark results

Here, we report on a numerical benchmark for an island-style FPGA using 22, 32, and 45 nm spin MOSFETs (spin FPGA) by improving standard benchmark tools.[35] Compared with other proposals,[36, 37] spin FPGA has an advantage because it is based on an Si transistor providing stable nonvolatile magnetic memory. A Monte Carlo simulation based on the Predictive Technology Model[38] is carried out to consider the variation of device size assuming fabrication difficulties. Although current experiments on MTJ[39] show that the maximum magnetocurrent (MC) ratio is 260% (RA~10 $\Omega \mu m^2$), in this paper we treat 100% $\leq$ MC ratio $\leq$ 1000% assuming future realization of larger MC. We model the spin MOSFET by changing the SPICE parameter (mobility) such that MC defined by MC = $(I_P-I_{AP})/I_{AP}$ coincide with a given MC ratio (I_P and I_{AP} are parallel and antiparallel currents, respectively.) For I_P, we use the same SPICE parameters as those of the conventional MOSFET. Area and speed of spin FPGA over 20 typical milliongate circuits are benchmarked with modified VPR ver.5 [35] for 22, 32, and 45 nm transistors. We take standard parameters such as Fs=3 (Wilton switch box), Fc-in=1.0 and Fc-out=0.25 with length 1 wire segment.[35] The average results over 200 Monte Carlo simulations for up to 20% (3 sigma) variations of length and width in 22 nm transistors, where the advantage of area, critical path delay and the area-delay product are defined by

$(\Theta_{cmos}-\Theta_{spin})/ \Theta_{spin}$ for Θ = {A (area), t_{delay} (critical path delay), A×t_{delay} (area-delay product)}.[11] The area-delay product is treated as a metric of FPGA performance. For the 22 nm transistor, a 16% average area reduction is realized.[11] This area reduction leads to a small critical path delay of circuits resulting in faster operation in spin FPGA. The speed is improved by an average of 24%. As the MC ratio increases, P/AP signals that go into an amplifier in spin LUT (Fig. 11) become clearer. This leads to more robust operation against the variation of transistors, resulting in a shorter delay. Thus, the area-delay product is improved, on average, by 43%. Figure 13 shows the summarized results of the bench-mark from 22 nm to 45 nm transistors. As mentioned above, as transistor scale decreases, the ratio of PMOS area to NMOS area decreases. This means that the effect of area reduction by spin MOSFET (NMOS) becomes greater resulting in better performance of small transistor nodes. One of the advantages of spin MOSFET compared with CMOS with the interlayer MRAM system is that, for spin MOSFET, the MC ratio change directly affects the sub-threshold region of MOSFET which leads to more efficient device operations. As the results of numerically benchmarked simulation for 22, 32, and 45 nm transistors, we find that the performance of spin FPGA becomes superior to that of conventional CMOS FPGA as transistor size decreases and MC ratio increases.

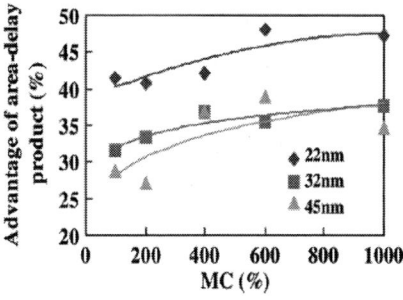

Fig. 13. A comparison of transistor generation. An average result of the benchmark calculation as a function of MC ratio.

7. Conclusions

Our detailed recent results for "spin injection, transport, and read/write operation in spin-based MOSFET" are reviewed and discussed. With a view to practical application, we proposed a novel spin-based MOSFET "Spin-Transfer-torque-Switching MOSFET (*STS*-MOSFET)" that offers non-volatile memory and transistor functions with CMOS compatibility, high endurance and fast write time. The *STS*-MOSFETs with Heusler alloy ($Co_2FeAl_{0.5}Si_{0.5}$) were prepared and reconfigurability of a novel spintronics-based MOSFET, *STS*-MOSFET was realized for the transport properties. The device showed *MC* and write characteristics with the endurance of over 10^5 cycles. The overall properties of the *STS*-MOSFETs show great potential for future reconfigurable integrated circuits based on CMOS technology. It was also clarified that the read characteristic can be improved in terms of *MC* ratio, however, is deteriorated in terms of the mobility by choosing connection configurations of the source and the drain in the *STS*-MOSFETs.

Moreover, spin FPGA was numerically benchmarked for 22, 32, and 45 nm transistors. We showed that the performance of spin FPGA becomes superior to that of conventional CMOS FPGA as transistor size decreases and MC ratio increases.

Acknowledgments

This work was partly supported by the New Energy and Industrial Technology Development Organization (NEDO) and the Grant-in-Aid for Scientific Research (B) (22360002) from JSPS.

References

1. International Technology Roadmap for Semiconductor, 2009 Edition, http://www.itrs.net/reports.html.
2. S. Sugahara and M. Tanaka, *Appl. Phys. Lett.* **84**, 2307 (2004).
3. T. Marukame, T. Inokuchi, M. Ishikawa, H. Sugiyama and Y. Saito, *IEDM*, 215 (2009).

4. T. Inokuchi, T. Marukame, T. Tanamoto, H. Sugiyama, M. Ishikawa and Y. Saito, *Symposium on VLSI Technology*, 119 (2010).
5. Y. Saito, T. Marukame, T. Inokuchi, M. Ishikawa, H. Sugiyama, T. Tanamoto, accepted to *Thin Solid Film* (Published online at http://dx.doi.org/10.1016/j.tsf.2011.03.073).
6. J. C. Slonczewski, *Phys. Rev. B* **71**, 024411 (2005).
7. W. H. Kautz, *IEEE Trans. Computers* 18, 719 (1969).
8. S. Datta and B. Das, *Appl. Phys. Lett.* **56**, 665 (1990).
9. T. Block, C. Felser, G. Jakob, J. Ensling, B. Muhling, P. Gutlich and R. Cava, *J. Solid State Chem.* **176**, 646 (2003).
10. K. Inomata, S. Okamura, R. Goto and N. Tezuka, *Jpn. J. Appl. Phys.* **42**, L419 (2003).
11. T. Tanamoto, H. Sugiyama, T. Inokuchi, T. Marukame, M. Ishikawa, K. Ikegami and Y. Saito, *J. Appl. Phys.* **109**, 07C312 (2011).
12. Y. Saito, T. Inokuchi, H. Sugiyama and K. Inomata, *Euro. Phys. J. B* **59**, 463 (2007).
13. X. Lou, C. Adelmann, S.A. Crooker, E.S. Garlid, J. Zhang, K.S.M. Reddy, S.D. Flexner, C. J. Palmstrom and P. A. Crowell: *Nature Phys.* **3**, 197 (2007).
14. A. Fert and H. Jaffrès, *Phys. Rev. B* 64, 184420 (2001).
15. A. T. Hanbicki, B. T. Jonker, G. Itskos, G. Kioseoglou and A. Petrou, *Appl. Phys. Lett.* **80**, 1240 (2002).
16. B. Huang, D. J. Monsma and I. Appelbaum, *Phys. Rev. Lett.* **99**, 177209 (2007).
17. O.M.J. van 't Erve, A.T. Hanbicki, M. Holub, C.H. Li, C. Awo-Affouda, P.E. Thompson and B.T. Jonker, *Appl. Phys. Lett.* **91**, 212109 (2007).
18. S. P. Dash, S. Sharma, R. S. Patel, M. P. de Jong and R. Jansen, *Nature* **462**, 491 (2009).
19. T. Sasaki, T. Oikawa, T. Suzuki, M. Shiraishi, Y. Suzuki and K. Noguchi, *Appl. Phys. Lett.* **96**, 12211 (2010).
20. T. Suzuki, T. Sasaki, T. Oikawa, M. Shiraishi, Y. Suzuki and K. Noguchi, *Appl. Phys. Express* **4**, 023003 (2011).
21. K. Inomata, N. Ikeda, N. Tezuka, R. Goto, S. Sugimoto, M. Wojcik and E. Jedryka, *Science and Technology of Advanced Materials* **9**, 014101 (2008).
22. N. Tezuka, N, Ikeda, S. Sugimoto and K. Inomata, *Jpn. J. Appl. Phys.* **46**, L454 (2007).
23. R.A. deGroot, F.M. Mueller, P.G. van Engen and K.H.J. Buschow, *Phys. Rev. Lett.* **50**, 2024 (1983).
24. D. Mauri, D. Scholl, H.C. Siegmann and E. Kay, *Phys. Rev. Lett.* **61**, 758 (1988).
25. A.H. MacDonald, T. Jungwirth and M. Kasner, *Phys. Rev. Lett.* **81**, 705 (1998).
26. C.H. Shang, J. Nowak, R. Jansen and J.S. Moodera, *Phys. Rev. B* **58**, R2917 (1998).
27. C. Hordequin, D. Ristoiu, L. Ranno and J. Pierre, *Eur. Phys. J. B* **16**, 287 (2000).
28. J. J. Attema, G. A. deWijs and R. A. deGroot, *J. Phys.: Condens. Matter* **19**, 315212 (2007).
29. Y. Sakuraba, M. Hattori, M. Oogane, Y. Ando, H. Kato, A. Sakuma, T. Miyazaki and H. Kubota, *Appl. Phys. Lett.* **88**, 192508 (2006).
30. A. Sakuma, Y. Toga and H. Tsuchiura, *J. Appl. Phys.* **105**, 07C910 (2009).
31. R. Shan, H. Sukegawa, W.H. Wang, M. Kodzuka, T. Furubayashi, T. Ohkubo, S. Mitani, K. Inomata and K. Hono, *Phys. Rev. Lett.* **102**, 246601 (2009).
32. Y. Shuto, R. Nakane, W. Wang, H. Sukegawa, S. Yamamoto, M. Tanaka, K. Inomata and S. Sugahara, *Appl. Phys. Exp.* **3**, 013003 (2010).
33. S.M.Sze and K. K. NG, *Physics of Semicnoductor Devices*, 3rd Edition, Ch. 6, p.293 (2007).
34. H. Sugiyama, T. Tanamoto, T. Marukame, M. Ishikawa, T. Inokuchi, and Y. Saito, *Int. Conf. Solid State Devices and Materials*, p.670 (2008).
35. V. Betz, J. Rose, and A. Marguardt, *Architecture and CAD for Deep-Submicron FPGAs*, Kluwer Academic Publishers, February 1999. ISBN 0–7923-8460–1. AQ2
36. A. DeHon, *ACM J. Emerging Technol. Comput. Syst.* **1**, 109 (2005).
37. C. Dong, D. Chen, S. Haruehanroengra, and W. Wang, *IEEE Trans. Circuits Syst.*, I: Regul. Pap. **54**, 2489 (2007).
38. W. Zhao and Y. Cao. http://www.eas.asu.edu/ptm/.
39. J. Hayakawa, S. Ikeda, F. Matsukura, H. Takahashi, and H. Ohno, *Jpn. J. Appl. Phys.* **44**, L587 (2005).

Flexible Thin Film Transistor Arrays as an Enabling Platform Technology:
Opportunities and Challenges

G. B. Raupp

Department of Electronic Engineering, City University of Hong Kong,
Kowloon, Hong Kong

Significant advances in large-area flexible electronics over the last
decade have created a tremendous opportunity for revolutionary
transformational engineered products and systems with unique and
desirable *form, fit and function*. In these "Flexible Systems",
materials and nano-, micro-, and macro-scale devices are
integrated to produce valuable multi-functional products that are
characteristically thin, lightweight, flexible, conformable, and
ultra-rugged for use under challenging conditions. The practical
opportunities seem limitless: one can envision new dual-use
technologies that can be deployed to address critical needs in a
diverse application space from health sciences to information and
decision technology, security and emergency response,
transportation, energy and the environment. Before this
compelling future is realized, however, major advances in three
critical areas must be achieved: (i) design and integration; (ii) flex-
compatible materials, structures and devices; and (iii) scaleable
and sustainable manufacturing processes.

Introduction

The tremendous success of the flat panel display industry can in large part be attributed to
the industry's ability to manufacture high quality large area thin film transistor (TFT)
arrays on glass substrates at high yield and throughput. Intensive strategic efforts over
the past decade to develop new processes and materials for manufacturing TFT arrays on
flexible plastic or metal foil substrates is pushing us to the dawn of a flexible display
industry.

In this paper we take a speculative look into the future, with the basic premise that
the ready manufacture of large area flexible TFT arrays will provide a tremendous
opportunity to create not just flexible displays, but revolutionary transformational
engineered flexible systems. These "Flexible Systems" (1) integrate sensors, actuators,
supporting electronics, and communication capability on versatile thin sheets that can
bend and fold like paper, stretch and relax like human skin, and integrate seamlessly with
their applications environment. Flexible Systems provide a natural physical interface
between microelectronics and the human scale. We envision new products such as smart
bandages that conform to the contours of the patient's body, monitor infection, prevent
pressure ulcers, deliver medications, and communicate with clinicians; large-scale skins
integrated into bridges, buildings, and aircraft that can measure strain throughout the
structure, harvest energy, and communicate wirelessly; and lightweight, wearable devices

that monitor physiological condition or human performance and communicate with the wearer through tactile means.

Flexible Systems release microelectronics from the arbitrary constraints of small size, brittle materials, and rigid, flat geometry. Sales of flexible systems could eventually exceed sales of VLSI circuitry. Analysts predict the flexible systems industry will grow to $100 billion in sales by 2020 (2) driven not by Moore's Law and the demand for greater computational power, but instead by market desire for products with greater functionality – products that can be simultaneously large-area, conformable, rugged, self-healing, ultra-thin, lightweight, and transparent.

To realize the full promise of Flexible Systems technology, fundamental and technical challenges in three critical inter-related areas must be addressed:

- *Standardized Industry-wide Systems-level Design Methods and Tools* -- system architecture guidelines, design rules, TCAD libraries, and integration strategies and methodologies.

- *Flex-compatible Functional Materials and Microstructures* -- high-performance, low-temperature TFTs and circuits along with critical transparent conductors and environmental barriers for energy harvesting, sensing, actuation, memory and wireless communication.

- *Scaleable and Sustainable Manufacturing* -- low temperature flex-compatible processes, flexible substrate handling protocols, and in the longer run novel additive and other environmentally-benign processing methods.

With an emphasis on TFT materials, devices and processes, this paper will summarize the ongoing efforts to effectively address these challenges and the R&D opportunities defined by the remaining principal technology gaps.

The Flexible Systems Opportunity and the State-of-the-Art

Technology Concepts

Figure 1 shows Flexible Systems technology concepts that are either actively under development or are under active consideration by industry and the academic community. The figure highlights the broad diversity of (i) *function* from sensing and diagnostic, to assistive and habitat adaptation, and to energy harvesting and energy efficiency; and (ii) *form* and *fit* from small to very large, from conformable and bendable, and from portable to wearable to building-, transportation-, and infrastructure-integrated systems.

In general the application will determine the technical and product-level requirements for a given opportunity. In an attempt to categorize the nearly limitless product and application opportunities, for purposes of discussion we focus on three areas as follows:

- In-body / On-body Systems for Human Health and Human Performance
- Smart Surfaces for Critical Infrastructure, Security, and Human Health
- Smart / Aware Spaces and Technology for the Human Experience

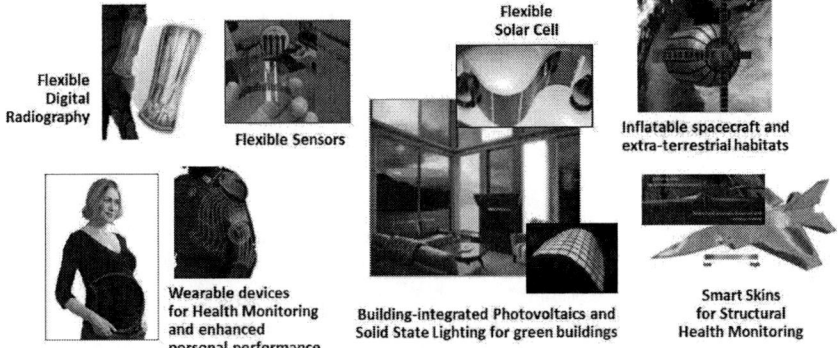

Figure 1. Flexible Systems Product and Applications Concepts

Table I summarizes representative high value applications in each area, and the requirements, features and functionality uniquely characteristic of each application type.

TABLE I. High Value Applications and Requirements for New Flexible Systems Technology

	Category 1: In-body / On-body Systems for Human Health and Human Performance	Category 2: Smart Surfaces for Critical Infrastructure, Security, and Human Health	Category 3: Smart / Aware Spaces and Technology for the Human Experience
Representative High Value Applications	Smart bandages for wound prevention and treatment. Implantable prosthetic devices and external muscle stimulators for rehabilitating spinal cord injury patients. Implantable neural prosthetic devices for quadriplegics. Wearable biomarker sensors for monitoring performance, stress, and fatigue. Smart hospital bed surfaces for wound care and avoidance.	Smart skins for large-area structural health monitoring. Flexible digital X-ray detectors for portable conformable digital radiography or see-through-the-wall technology. Smart skins for aircraft that tolerate extreme temperature. Inexpensive large-area pressure-sensitive pads that track movement of people and objects.	Smart surfaces and wearables to assist the visually impaired. Smart hospital / extended care / rehabilitation spaces. Adaptive and customizable learning / work environments. Sensing and actuation for energy-saving climate control within buildings. Inflatable structures for human habitats in solar system exploration.
Characteristic Requirements, Features and Functionality	Small, stretchable, compliant, soft, biocompatible. Able to sense physiological phenomena. Self-powered with electrical stimulation capability and short-range wireless interrogation. Integrated memory. Automated delivery of pharmaceuticals.	Stretchable, compliant, low deformation, environmentally robust. Moderate to very large area. Ability to sense environmental conditions and mechanical strain , and to identify long-term trends. Self-powered with capability for short-range wireless communication / interrogation.	Conformable, seamlessly integrated, inobtrusive. Small wearable to large-area building-integrated. Ability to sense human health, activity, and environmental. Integrated actuation. Highly networked.

State of the Art of Flexible Systems

Flexible RFID tags have entered the technological mainstream. Several companies have begun to offer flexible photovoltaic systems. It is anticipated that manufacturers of e-readers will soon introduce products that contain flexible displays. However, no manufacturing company has as yet scheduled the introduction of another type of product based on flexible electronics.

Figure 2 provides a non-exhaustive but representative snapshot of the state of the art of flexible electronic devices in academic laboratories. Shown are (a) a rubber-like stretchable active transistor matrix with elastic conductors (3); (b) stretchable electrical conductors on a flexible elastomeric substrate (4); (c) an organic thin film transistor active matrix backplane (5); (d) an "electronic eye" camera made by transferring silicon photodiodes onto a flexible elastomer (6); (e) a non-volatile memory array (7); and (f) arrays of thin film transistors on a large area transparent flexible plastic substrate (8). All laboratory prototypes built to date are single-function devices. No one has developed procedures or methodologies for integrating multiple functions into the same flexible system. Within the university community, the prototypes and the detailed fabrication and especially the integration techniques differ from lab to lab, and the community has yet to adopt standards for design rules, integration strategies and tool kits.

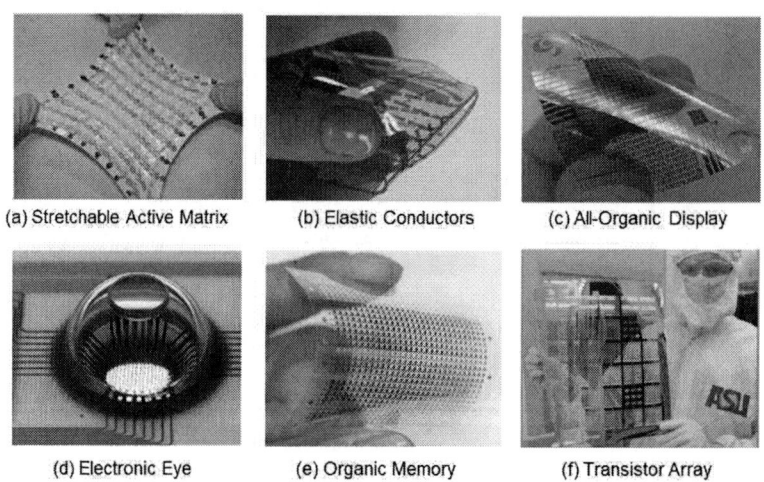

(a) Stretchable Active Matrix   (b) Elastic Conductors   (c) All-Organic Display

(d) Electronic Eye   (e) Organic Memory   (f) Transistor Array

Figure 2. Flexible Electronics State of the Art Devices in Academic Laboratories

Principal Challenges for the Technical Community

Challenge 1: Standardized Industry-wide Systems-level Design Methods and Tools

With the ad hoc, customized nature of the creation of today's flexible electronic systems, the pathway to standards and a standardized approach is not yet clear. It is however tempting to speculate that the community may follow a path that mirrors that which evolved in the flat panel display industry, in which large area TFT array active matrix backplanes are manufactured in fabs that are substantially different from, and physically separated from, integration with their respective imaging frontplanes (LCD, OLED or EPD). In this construct, a generic large-area flexible electronic system would consist of "backplane" electronics on a flexible substrate, a "functional frontplane," and encapsulation packaging. All components would require the degree of flexibility demanded by the product/application. The frontplane would contain the sensors, actuators, energy harvesters, or other devices that interact with the external world. The backplane TFT array and other circuitry would control the activity of the frontplane functions and provide communication and intelligence. The integrated system could be fabricated monolithically by sequential bottom-up processing as in OLED display module manufacturing, or might be fabricated separately and then laminated together as in EPD module manufacturing.

The flexible displays that will soon reach the market in e-readers exemplify the simplicity of this latter approach, which reduces the complexity of processes and materials and simplifies the distribution of development tasks. Moreover, flexible frontplane and flexible backplane technology development can be effectively pursued in parallel, and in the case of certain frontplane technologies may be most cost-effectively achieved with roll-to-roll processing. Technical issues to be addressed with a "divide and conquer" lamination strategy include adhesion, alignment, and reliable electrical coupling between the frontplane and backplane. By comparison, the monolithic direct fabrication approach may provide improved system reliability, frontplane-backplane coupling, and alignment accuracy, but is more demanding in terms of constraints on materials and processes due to the low-temperature processing required by plastic substrates.

Major generic architectural issues to be addressed include the mode of short-distance electrical interfacing between layers (contact *vs.* contactless) and associated alignment approaches, and long-distance communication in the backplane (wired *vs.* wireless). Once standard approaches and hardware solutions are developed and adopted, the community will need to develop design rules, design libraries, and technology CAD packages that will enable application engineers to rapidly and effectively design their own flexible system prototypes and products.

Challenge 2: Flex-compatible Materials and Microstructures

Several candidate material technologies have emerged for thin film transistors on flexible substrates, including thin film silicon, metal oxides, organic semiconductors, and carbon nanotubes (CNT). Amorphous silicon (*a-Si:H*) has already entered the technological mainstream; it underpins the $100B flat panel display industry. Organic semiconductors are under active manufacturing development, but their performance is substantially poorer than *a-Si:H*. CNTs suffer from purity limitations and in the context of highly uniform, large area TFT arrays, are substantially more difficult to fabricate. Thus a logical technology roadmap to pursue includes *a-Si:H* and other forms of silicon in the short term and higher performance oxide TFTs in the medium term, since both material sets could leverage the substantial industrial infrastructure and supply network already in place in the FPD industry. In the longer term, even higher performance revolutionary materials such as CNTs or silicon nano-wires could be pursued.

Silicon-based TFT Technologies. Thin film silicon technologies include *a-S:H*, with typical laboratory-scale carrier mobility of about ~1 cm^2/V-s, and *nano- or micro-crystalline Si*, with carrier mobility up to ~10 cm^2/V-s. In production, tradeoffs between performance and processability/tact time generally lead to somewhat lower values (*e.g.* typical *a-Si:H* mobility in an FPD fab is 0.4 cm^2/V-s). Because of the FPD industry move away from aggressive development of laser crystallized low temperature poly-silicon and the special problems introduced by laser processing on plastic, this technology is not considered a serious candidate for flexible electronics.

The principal challenge for fabrication of *a-Si:H* TFTs for flexible systems is identification of low temperature processing windows that are compatible with the flexible substrates and yet produce TFTs of equivalent or better performance than those fabricated in a manufacturing environment on glass at 300-380 °C. The most critical steps are the active stack deposition processes in which silane-sourced plasma-enhanced chemical vapor deposition (PECVD) is the industry standard. For typical plastic substrates these windows must employ temperatures below 200 °C or even lower for best processability. If the process temperature is reduced without substantial modifications to other conditions in the process, the quality and performance of the active device stack is significantly compromised; specifically the TFTs exhibit higher threshold voltage, lower saturation mobility, greater electrical stress-induced threshold voltage shift (instability), and greater contact resistance and turn-on voltage offset (9-11).

It is now well known that the *a-Si:H* silicon hydride concentrations can be effectively controlled through dilution of the feed gas with hydrogen, with a trade-off in slower deposition rates and longer tact times. With appropriate choice of feed gas composition and flow rate, total pressure and plasma density conditions, large area TFT arrays on plastic can be fabricated to give statistically averaged μ_{sat} equal to 0.9 cm^2/V-s with ON/OFF ratios greater than 10^8, and with TFT yields in the 99-100% range in a manufacturing pilot line environment at temperatures ≤180 °C (12). In all ways the performance metrics of these TFTs on plastic meet or exceed those fabricated on glass in the FPD industry, with one key critical exception. Like the commercial TFTs which are inherently unstable, the threshold voltage V_{th} increases gradually with a DC voltage applied continuously to the gate, but the degradation is greater for the lower temperature processes designed for flexible substrates (13). The two accepted mechanisms responsible for the ΔV_{th} in a-Si:H TFTs are charge injection in the silicon nitride (SiN_x) gate insulator and creation of defect states in the a-Si:H conducting channel (14). Field-effect experiments have provided evidence that mobile carriers are responsible for breaking the weak Si–H bonds resulting in the creation of charged defect states (dangling bonds). By extending techniques used to localize hot electron degradation in MOSFETs, experiments at Arizona State University have localized the degradation of *a-Si:H* to the gate dielectric- channel interface (15).

Wagner and Sturm's group at Princeton University have recently demonstrated remarkably stable nano-/micro-crystalline silicon TFTs (16) that indicate promise for eventually being able to fabricate stable silicon-based TFTs on low temperature plastics. Using a combination of stress control and *in situ* etching of hydrogen bonds, they were able to modify the surface of the gate dielectric on which the silicon was deposited, thereby increasing the tendency of the silicon to order and significantly reducing dielectric and channel defectivity. Figure 3 compares extrapolated lifetime tests for industry standard TFTs and a sequence of successively improved devices; the sum total of the improvements produces TFTs with unprecedented lifetimes of ~100 years, 1000 times longer than current industry standards.

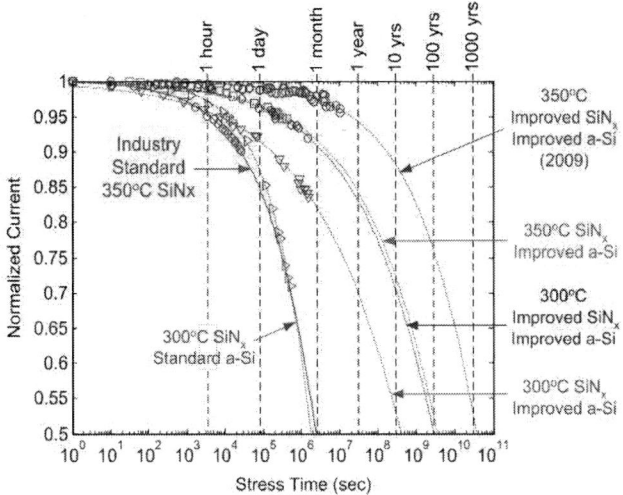

Figure 3. Extrapolated lifetime measurements for TFTs produced under industry standard conditions and with successive improvements to gate dielectric and channel layers. (16)

Metal Oxide Technologies. Transparent oxide semiconductors such as ZnO and mixed oxides of gallium, indium, zinc, aluminum, hafnium, and tin promise carrier mobility up to ~100 cm^2 / V-s (17-19). With a wide bandgap of about 3.4 eV, they are transparent in the visible region, and are therefore more stable under visible light exposure. These oxides can be deposited using conventional RF (20) or DC (21) sputtering techniques, as well as other less conventional techniques near room temperature, which in principle would enable their fabrication on flexible plastic substrates. Unfortunately the as-deposited materials typically suffer from significant threshold voltage instability. Stabilization of the threshold voltage can be achieved through one or more high temperature (>300 °C), post processing steps (22). There is therefore substantial research and development underway to identify alternative low temperature deposition process confditions and/or novel device architectures that will provide high performance, stable oxide TFTs that are flex-compatible.

The number of candidate alternatives is vast, and this situation provides both a major challenge for the industrial community and a major opportunity for the academic/research community. Figure 4 highlights the complexity of the problem by visually summarizing *some* of the options. The oxide active channel layer may be pure ZnO, or a binary, ternary or even quarternary mixed oxide including Ga, In, Sn, Al, and others. At this stage the optimal pairing of the various channel layer options with the best respective gate dielectric layer has not been established. Thus the number of potential pairs of materials for the channel and the gate dielectric is large. Moreover, candidate

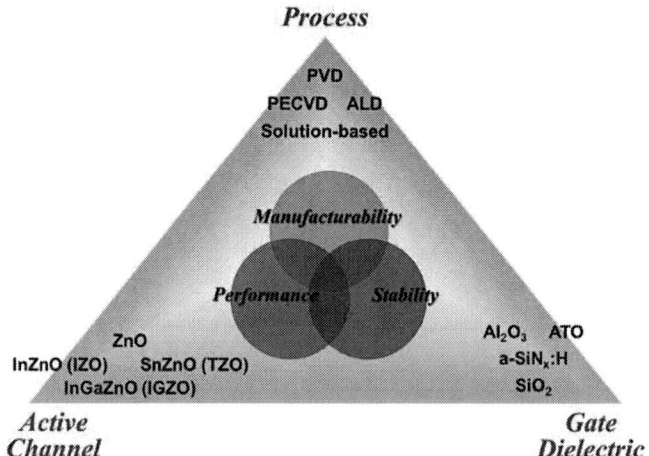

Figure 4. Graphic highlighting the large number of possible combinations of process type, active channel and dielectric layer choices for low temperature oxide TFTs.

low temperature process techniques for the two layers inlcude not only sputtering but other physical vapor depositon (PVD), chemical vapor deposition methods including atomic layer deposition (ALD) methods, and solution-based processing methods such as sol-gel. The academic community could contribute significantly to the challenge of finding the best combination of channel, gate dielectric and process that optimizes performance, stability and manufacturability through fundamental studies of the interactions of the channel and gate material to provide logical pairing guidelines, of the mechanistic sources of instability to provide pathway for stability enhancement, and of thin film deposition to provide process models for process window identification.

Challenge 3: Scaleable and Sustainable Manufacturing Processes

Although it is tempting to speculate on potential revolutionary manufacturing technologies including roll-to-roll substrate handling and direct printing that would dramatically alter the competitive landscape, the fastest and least capital-intensive way for flexible electronic systems to enter the mainstream may be through effective leveraging of the tremendous historical investment in TFT manufacturing infrastructure for commercial glass-based flat panel displays. In this context many groups are pursuing new handling protocols aimed at enabling high quality TFT arrays to be directly fabricated on flexible substrates in automated manufacturing tools that are built to process rigid glass substrates.

Figure 5 depicts three competing approaches that are under active development at the pilot line or manufacturing scale. From left-to-right these are temporary bonding and debonding, coat and laser release, and fabricate with layer transfer. In the temporary bonding / de-bonding approach developed by the Flexible Display Center at Arizona State University (12), ITRI in Taiwan as well as others, a flexible substrate is temporarily

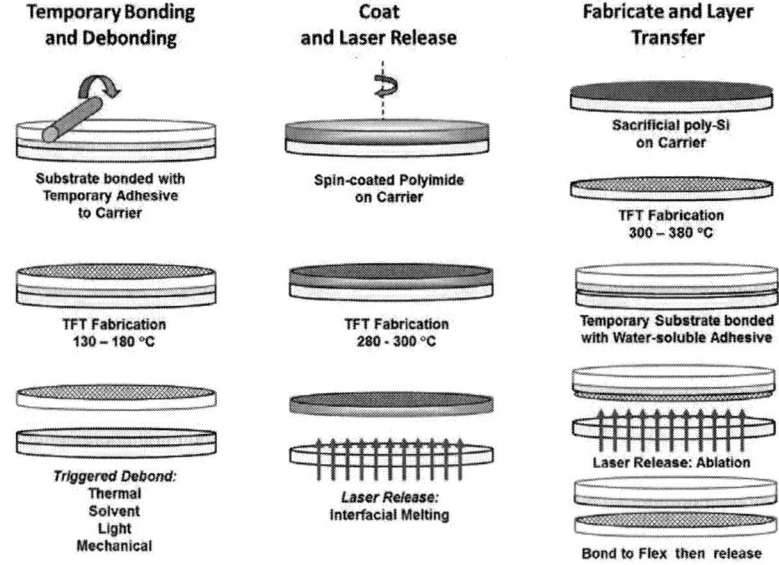

Figure 5. Flexible substrate handling protocols under manufacturing development

adhered to a rigid carrier plate with a releasable adhesive, and the carrier-adhesive-substrate system is then processed using standard automated TFT fabrication tools. The rigid carrier gives the structural support and suppresses deformation of the flexible substrate during processing. Following full TFT array fabrication the flexible substrate is released from the carrier through any of a number of triggered release mechanisms including mechanical, solvent, ultraviolet light or thermal release processes.

In the coat – laser release process pioneered by IBM a thin polymer layer is cast from solution (polyimide spin-coating is a common approach), followed by microelectronics fabrication and backside excimer laser-induced release by melting/ablation of the polyimide at the glass-polymer interface (23-24). Philips has developed a version of this process known as EPLaR™ (electronics on plastic by laser release) to produce TFT arrays for reflective flexible displays (25-26); the technology was licensed to Prime View International (since renamed E Ink Holdings) who is actively advancing its further development for flexible display commercialization. In the layer transfer process, TFT arrays are fabricated directly on glass and then laser-released and transferred to a flexible substrate. Seiko-Epson has pioneered this approach to produce poly-silicon TFT arrays with a process they have trademarked as SUFTLA™ (Surface Free Technology by Laser Annealing/Ablation) (27-28). Each of the three processes has unique inherent advantages, challenges and limitations as is summarized in Table II. In the context that the work to date has focused on GEN I and GEN II pilot and manufacturing scale equipment, a key outstanding unanswered question is the ultimate scale-ability of each of these process protocols. In this context the temporary bonding protocol is described in greater detail in the paragraphs which follow.

Table II. Capability and limitations comparison of flexible electronics manufacturing protocols

	Temporary Bonding and Debonding	*Coat and Laser Release (EPLaR™)*	*Fabricate and Layer Transfer (SUFTLA™)*
Flexible substrate	High surface quality polymer or metal foil	Solution castable polymers (Polyimide, BCB)	Any
TFT Process Temperature Limit	Substrate dependent (180 °C for HS-PEN, much higher for Stainless Steel and other metal foils)	Polymer-dependent (280 °C for Polyimide)	Typical glass-based TFT limits
Flexible Substrate Distortion	Can be significant – but can be controlled to negligible level	Negligible	Not applicable since fabrication is on glass
Release Process	Rapid, automated, dry	Laser interfacial melting	Laser ablation of sacrificial layer
Scalability	?	?	?

Successful implementation of the temporary bonding protocol required simultaneous development of new custom carriers and temporary adhesive materials, adapted manufacturing tools for automated bonding and debonding, and development of new robust processes and handling protocols. A key to successful demonstration of this technology was a systems-level methodology that considered the fundamental thermo-mechanical interactions of the substrate carrier-adhesive-substrate system. A second key was the development of custom high performance temporary adhesives tailored for the plastic substrates and TFT process temperatures.

The most significant issue encountered with this approach is the stress that is developed during the bonding-debonding processes as well as during the TFT direct fabrication process steps. In the higher temperature steps in the process flow (typically the active stack steps), the thermal property mismatches between the carrier, adhesive, and flexible substrate become important. The thermal property mismatch induced stresses lead to bowing of the carrier-adhesive-substrate system, which can lead to wafer handling problems in processing equipment, delamination of the flexible substrate from the rigid carrier, or substrate distortion that leads to layer alignment problems. The choice of alumina as a carrier provides a non-obvious custom solution through a more closely matched coefficient of thermal expansion with the plastic substrate. Properties of the adhesives used to bond the flexible substrate to the rigid carrier were found to have a significant effect on the bow of the bonded system (29-30). Specifically, the relative viscoelastic flow properties of the bonding adhesive to that of the bonded flexible substrate is directly correlated to bow and distortion. When the loss factor of the adhesive is less than that for the plastic substrate, precise registration of layers during photolithography is observed. In summary the properties of the rigid carrier, flexible substrate, and adhesive must all be considered in the selection of a bonded flexible substrate system. As the protocol is scaled to larger generations of manufacturing equipment, sound fundamental models describing the interactions could provide valuable scaling laws and insights into fruitful direct development paths.

Conclusions

Significant advances in large-area flexible electronics over the last decade have positioned us to realize an exciting new world in which Flexible Systems technologies and consumer products have far-reaching impact on our daily lives and the way we intrinsically interface with and use electronics-based technologies. Large area, high performance (equivalent to today's *a-Si:H* on glass or better) flexible TFT arrays are seen as the key enabler. However, there remains significant work to be done if this promise is to be realized. Today's flexible electronics R&D community lacks a focus on system-level applications beyond flexible displays. No one has created design tools that allow cost-effective, scalable integration of multiple functions in a single system. Large area flexible TFT arrays are not yet routinely manufactured at low cost. We lack widely-accepted processing protocols and the dedicated manufacturing infrastructure supply chain. Nonetheless, the challenges are well understood at this point in the technology evolution, and large-scale, focused efforts on the part of industry, government and academia can close the fundamental gaps and produce the technology enablers that are needed.

Acknowledgments

Many of the ideas and positions presented in this paper were developed over the last few years in multiple detailed discussions and interactions with some of the pioneers and key thought leaders in the field of flexible electronics, including Sigurd Wagner and Jim Sturm at Princeton University, Bruce Gnade at University of Texas at Dallas, Takao Someya at University of Tokyo, and David Allee, Jiping He, Sethuraman "Panch" Panchanathan and Bryan Vogt at Arizona State University; I have been fortunate to benefit from their wisdom, creativity and friendship. I also would like to thank my friends and consultants Donn Forbes and Ben Rowland for working with us to help crystallize our thinking and effectively articulate our ideas.

References

1. D. Gamota, "Near-term Opportunities for Large Area Flexible Electronics", *Circuits & Assembly*, April 14, 2009.
2. Nanomarkets, "Printable Electronics: Roadmaps, Markets and Opportunities", Sept. 19, 2006. *Also:* IDTechEx, "Organic Electronics Forecasts, Players, Opportunities 2005-2025", June 17, 2005.
3. Tsuyoshi Sekitani, Yoshiaki Noguchi, Kenji Hata, Takanori Fukushima, Takuzo Aida, and Takao Someya, *Science*, **321**, 1468 (2008).
4. S. Wagner, S.P. Lacour, J. Jones, P.I. Hsu, J.C. Sturm, T. Li, and Z. Suo, *Physica E*, **25**, 326 (2004).

5. G. Gutierrez-Heredia, L.A. Gonzalez, D. Berman, A. Avendano, U. Bhansali, H.N. Alshareef, B.E. Gnade and M. Quevedo-Lopez, *Mat. Sci. Engineering: B*, in press.
6. H.C. Ko, M.P. Stoykovich, J. Song, V. Malyarchuk, W.M. Choi, C.-J. Yu, J.B. Geddes III, J. Xiao, S. Wang, Y Huang and J.A. Rogers, *Nature*, **454**, 748 (2008).
7. T. Sekitani, T. Yokota, U. Zschieschang, H. Klauk, S. Bauer, K. Takeuchi, M. Takamiya, T. Sakurai and T. Someya, *Science*, **326**, 1516 (2009).
8. Flexible Display Center at Arizona State University, unpublished.
9. M. J. Powell, *IEEE Trans. Elec. Dev.*, **36**(12), 2753 (1989).
10. C.-S. Yang, L. L. Smith, C. B. Arthur, and G. N. Parsons, *J. Vac. Sci. Technol. B*, **18(2)**, 683 (2000).
11. K. Long, H. Gleskova, S. Wagner, and J. C. Sturm, *Proc. 62nd Device Res. Conf.*, 89 (2004).
12. G.B. Raupp, S.M. O'Rourke, C. Moyer, B.P. O'Brien, S.K. Ageno, D.E. Loy, E.J. Bawolek, D.R. Allee, S.M. Venugopal, J. Kaminski, D. Bottesch, J. Dailey, K. Long, M. Marrs, N.R. Munizza, H. Haverinen, and N. Colaneri, *JSID* **15**(7), 445 (2007).
13. H. Gleskova and S. Wagner, *IEEE Trans. Electron Dev.*, **48**, 1667 (2001).
14. M.J. Powell, **36**, 2753 (1989).
15. R. Shringarpure, S. Venugopal, L.T. Clark, D.R. Allee and E. Bawolek, *IEEE Elect. Dev. Lett.*, **29**, 93 (2008).
16. B. Hekmatshoar, S. Wagner and J.C. Sturm, *Appl. Phys. Lett.*, **95**, 143504 (2009).
17. H. Hosono, M. Yasukawa and H. Kawazoe, *J. Non-Cryst. Solids*, **203**, 334 (1996).
18. N. Itagaki, T. Iwasaki, H. Kumomi, T. Den, K. Nomura, T. Kamiya and H. Hosono, *Phys. Stat. Sol. (a)*, **205**(8), 1915 (2008).
19. H. Q. Chiang, J. F. Wager, R. L. Hoffman, J. Jeong, and D. A. Keszler, *Appl. Phys. Lett.*, **86**, 013503 (2005).
20. J. S. Park, J. K. Jeong, Y. G. Mo, H. D. Kim and S.I. Kim, *Appl. Phys. Lett.*, **90**, 262106 (2007).
21. N. Ito, Y. Sato, P. K. Song, A. Kaijio, K. Inoue and Y. Shigesato, *Thin Solid Films* **496**, 99 (2006).
22. H. Q. Chiang, B. R. McFarlane, D. Hong, R. E. Presley and J. F. Wager, *J. Non-Cryst. Solids*, **354**, 2826 (2008).
23. G. Arjavalingam, A. Deutsch, F. E. Doany, B. K. Furman, D. J. Hunt, C. Narayan, M. M. Oprysko, S. Purushothaman, V. Ranieri, S. Renick, J. M. Shaw, J. S. Wilczynski, and D. F. Witman, U.S. Patent 5,258,236, Nov. 2, 1993.
24. F. E. Doany and C. Narayan, *IBM J. Res. Dev.*, **41**(1/2), 151 (1997).
25. I. French and D. McCulloch, International Patent WO 2005/050754 A1, June 2, 2005.
26. H. Lifka, C. Tanase, D. McCulloch, P. van de Weijer, and I. French., *SID Symp. Digest Tech. Papers,* **38**(1), 1599 (2007).
27. T. Hashimoto, A. Takakuwa, T. Kamakura, and S. Utsunomiya, U.S. Patent 7,029,960, April 18, 2006.
28. M. Miyasaka, *JSID*, **15**(7), 479 (2007).
29. J. Haq, S. Ageno, G.B. Raupp, B.D. Vogt and D. Loy, *J. Appl. Phys.* **108**, 114917 (2010).
30. J. Haq, B.D. Vogt, G.B. Raupp and D. Loy, *Microelectronic Engineering* (2011) in press. doi:10.1016/j.mee.2011.02.099

Author Index

Amano, S.	97	Kuo, Y.	157
Asano, T.	15	Kwok, H.	23, 49
Baiano, A.	65	Larcher, L.	189
Beenakker, C.	65	Lee, H.	167
Belarbi, K.	39	Lhermite, H.	39
Bersuker, G.	189	Li, Z.	115
Bonnaud, O.	29	Lin, Z.	115
Brotherton, S. D.	3		
		Ma, T.	207
Chen, T.	65	Magyari-Köpe, B.	167
Coulon, N.	39	Maiolo, L.	3
Cuscunà, M.	3	Mariucci, L.	3
		Meng, Z.	49
Du, J.	133	Miyake, H.	89, 97
		Mofrad, M. R.	65
Feng, L.	105	Mohammed-Brahim, T.	39
Fonash, S.	141	Moon, K.	115
Fortunato, G.	3	Murota, J.	181
Furutani, K.	77		
		Nakagawa, G.	15
Garg, P.	141	Nihei, M.	121
Guo, X.	57, 105	Nishi, T.	89, 97
		Nishi, Y.	167
Harada, N.	121		
Hayashi, K.	121	Okazaki, K.	97
Hirakata, Y.	89, 97		
Ho, T.	49	Padovani, A.	189
		Pan, S.	141
Ishihara, R.	65	Park, S.	167
Ivanov, E.	199	Pecora, A.	3
Jacques, E.	29	Rapisarda, M.	3
		Raupp, G. B.	229
Kandoussi, K.	39	Roca i Cabarrocas, P.	147
Kaneyasu, M.	89	Rogel, R.	29
Kato, K.	77		
Kondo, D.	121	Saito, Y.	217
Koyama, J.	77, 89, 97	Sakakura, M.	89, 97

Sakuraba, M.	181
Sato, R.	89
Sato, S.	121
Scott, D.	141
Sekine, Y.	77
Shannon, J. M.	57
Shionoiri, Y.	77
Silva, S. R.	57
Simon, C.	39
Sporea, R. A.	57
Sun, P.	49
Sun, X.	207
Tillack, B.	181
Valletta, A.	3
Vandelli, L.	189
Wang, C.	133
Wang, X.	133
Winter, C.	141
Wong, C. P.	115
Wong, M.	23
Wong, M.	49
Wu, J.	141
Xie, W.	133
Xu, J.	133
Xu, X.	105
Yagi, K.	121
Yamada, A.	121
Yamazaki, S.	77, 89, 97
Yao, Y.	115
Yokoyama, N.	121
Yu, L.	147
Yuan, Y.	199
Zhang, P.	29
Zhao, S.	23, 49
Zhou, W.	23